Leckie ✕ **Leckie**

Scotland's leading educational publishers

Higher
BIOLOGY
GRADE BOOSTER

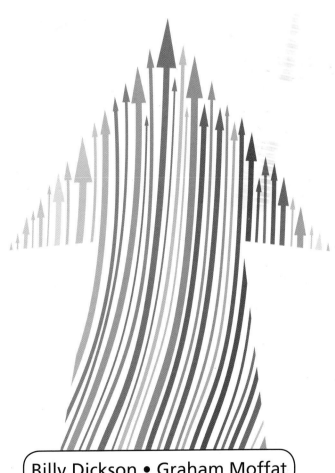

Billy Dickson • Graham Moffat

Higher BIOLOGY GRADE BOOSTER

© 2019 Leckie & Leckie Ltd

001/10012019

10 9 8 7 6 5 4 3 2 1

ISBN 9780007590834

Published by
Leckie & Leckie Ltd
An imprint of HarperCollinsPublishers
Westerhill Road, Bishopbriggs, Glasgow, G64 2QT
T: 0844 576 8126   F: 0844 576 8131
leckieandleckie@harpercollins.co.uk
www.leckieandleckie.co.uk

Commissioning editor: Kerry Ferguson
Managing editor: Craig Balfour

Special thanks to
Jess White (copyedit)
Jan Schubert (proofread)
Lauren Reid (proofread)
Jouve India (layout, illustration and project management)

Printed and bound in China by RR Donnelley APS

A CIP Catalogue record for this book is available from the British Library.

Acknowledgements
All images © Shutterstock.com

Chapter 25, page 32 and 33, SQA questions reproduced with permission (solutions do not emanate from SQA), Copyright © Scottish Qualifications Authority.

Contents

Progress grid		*4*
Introduction		8

Section 1 DNA and the genome

Chapter 1	Structure and organisation of DNA	11
Chapter 2	Replication of DNA	17
Chapter 3	Control of gene expression	24
Chapter 4	Cellular differentiation	35
Chapter 5	The structure of the genome	41
Chapter 6	Mutation	43
Chapter 7	Evolution	51
Chapter 8	Genomic sequencing	59

Section 2 Metabolism and survival

Chapter 9	Metabolic pathways	66
Chapter 10	Cellular respiration	75
Chapter 11	Metabolic rate	85
Chapter 12	Metabolism in conformers and regulators	90
Chapter 13	Metabolism and adverse conditions	99
Chapter 14	Environmental control of metabolism in microorganisms	105
Chapter 15	Genetic control of metabolism in microorganisms	113

Section 3 Sustainability and interdependence

Chapter 16	Food supply, plant growth and productivity	120
Chapter 17	Plant and animal breeding	130
Chapter 18	Crop protection	139
Chapter 19	Animal welfare	148
Chapter 20	Symbiosis	151
Chapter 21	Social behaviour	155
Chapter 22	Components of biodiversity	164
Chapter 23	Threats to biodiversity	167

Section 4 Assignment

Chapter 24	Assignment	174

Section 5 Skills of scientific inquiry

Chapter 25	Skills of scientific inquiry	

Skills of scientific inquiry material can be downloaded from the Leckie and Leckie website and unlocked using the **password: codon123**
https://collins.co.uk/pages/scottish-curriculum-free-resources

Flashcard glossary

Flashcards can be downloaded from the Leckie and Leckie website:
https://collins.co.uk/pages/scottish-curriculum-free-resources

Progress grid

Chapter	Area	Page	Content	Questions	Mark	Check (✓)
1	Section 1: DNA and the genome	11	Structure and organisation of DNA	C-type questions	/46	
				A-type questions		
				Extended response questions		
2		17	Replication of DNA	C-type questions	/41	
				A-type questions		
				Extended response questions		
3		24	Control of gene expression	C-type questions	/67	
				A-type questions		
				Extended response questions		
4		35	Cellular differentiation	C-type questions	/28	
				A-type questions		
				Extended response questions		
5		41	The structure of the genome	C-type questions	/11	
				A-type questions		
				Extended response questions		
6		43	Mutation	C-type questions	/41	
				A-type questions		
				Extended response questions		
7		51	Evolution	C-type questions	/45	
				A-type questions		
				Extended response questions		
8		59	Genomic sequencing	C-type questions	/33	
				A-type questions		
				Extended response questions		

Chapter	Area	Page	Content	Questions	Mark	Check (✓)
9		66	Metabolic pathways	C-type questions	/65	
				A-type questions		
				Extended response questions		
10		75	Cellular respiration	C-type questions	/76	
				A-type questions		
				Extended response questions		
11		85	Metabolic rate	C-type questions	/28	
				A-type questions		
				Extended response questions		
12	Section 2: Metabolism and survival	90	Metabolism in conformers and regulators	C-type questions	/64	
				A-type questions		
				Extended response questions		
13		99	Metabolism and adverse conditions	C-type questions	/42	
				A-type questions		
				Extended response questions		
14		105	Environmental control of metabolism in microorganisms	C-type questions	/42	
				A-type questions		
				Extended response questions		
15		113	Genetic control of metabolism in microorganisms	C-type questions	/38	
				A-type questions		
				Extended response questions		

Chapter	Area	Page	Content	Questions	Mark	Check (✓)
16		120	Food supply, plant growth and productivity	C-type questions	/76	
				A-type questions		
				Extended response questions		
17		130	Plant and animal breeding	C-type questions	/40	
				A-type questions		
				Extended response questions		
18		139	Crop protection	C-type questions	/69	
				A-type questions		
				Extended response questions		
19	Section 3: Sustainability and interdependence	148	Animal welfare	C-type questions	/15	
				A-type questions		
				Extended response questions		
20		151	Symbiosis	C-type questions	/24	
				A-type questions		
				Extended response questions		
21		155	Social behaviour	C-type questions	/50	
				A-type questions		
				Extended response questions		
22		164	Components of biodiversity	C-type questions	/18	
				A-type questions		
				Extended response questions		
23		167	Threats to biodiversity	C-type questions	/39	
				A-type questions		
				Extended response questions		

Chapter	Area	Page	Content	Check (✓)
24	Section 4: Assignment	174	Introduction	
		174	Research stage	
		174	Report stage	
		174	Mark allocations	
		175	Tips on challenging aspects of the assignment	

Chapter	Area	Page	Content		Questions	Mark	Check (✓)
25	Section 5: Skills of scientific inquiry	1	Planning				
		8	Evaluating				
		9	Selecting				
		17	Presenting				
		18	Processing				
		24	Concluding				
		25	Predicting				
		27	Examples of large skills questions	Large experimental questions	DNA and the genome	/9	
					Metabolism and survival	/8	
					Sustainability and interdependence	/8	
				Large data questions	DNA and the genome	/7	
					Metabolism and survival	/7	
					Sustainability and interdependence	/8	

Skills of scientific inquiry material can be downloaded from the Leckie and Leckie website and unlocked using the **password: codon123**

https://collins.co.uk/pages/scottish-curriculum-free-resources

Introduction

Welcome to your Higher Biology Grade Booster. Biology is a challenging Higher with lots of content and a wide range of scientific skills to master. Using this book can help you meet the challenge.

Higher Biology course organisation and assessment

Higher Biology is **organised** into three main biological areas:

- DNA and the genome
- Metabolism and survival
- Sustainability and interdependence

Each area has a specific body of knowledge which you need to able to demonstrate and apply, and provides contexts to develop the skills of scientific inquiry.

The course has three **assessment** components:

- Assignment (20 marks – scaled to 30 marks)
- Course examination Paper 1 (25 marks)
- Course examination Paper 2 (95 marks)

The assignment

The assignment is set by the Scottish Qualifications Authority (SQA). It is undertaken in class and is divided into two stages:

Stage 1: Research stage – collecting data by carrying out an experiment and internet or literature search

Stage 2: Report stage – writing a report on your research

The examination

The examination is set by SQA and has two Papers.

Paper 1: 25 multiple choice items (25 marks)

Paper 2: Structured questions and extended response questions (ERQs) (95 marks)

There will be two ERQs totalling 9–15 marks, a large practical question for 5–9 marks and a large data question for 5–9 marks included.

What you should already know

Higher Biology builds on much of the content and skills in National 5 Biology. In preparing this book, we have assumed a basic knowledge of the content and skills of National 5.

What this book is for

The aim of this book is to guide you to increase your scores in tests and exams and so lead to a better grade for your Higher Biology. It is not a textbook in the usual way. It allows you to revise all aspects of your course through attempting questions and so lets you identify the areas that need most work.

Structure of this book

Sections 1–3 each cover an area of the Higher Biology course. Each area is broken down into the individual key areas of the course, which are each given a chapter.

Section 4 deals with the assignment and provides help in maximising marks for this component of your Higher Biology course. We also use the symbol ▌, as described on page 10, in this section.

Section 5 covers skills of scientific inquiry. Pages for this section can be downloaded from the Leckie and Leckie website and unlocked using the **password: codon123**. **https://collins.co.uk/pages/scottish-curriculum-free-resources**. It is broken down into the seven main skill areas and aims to help by providing methods for tackling these questions. We use ▌ and ➡ symbols, as described below, in this section too. There is also a set of examples of large practical and data questions with model answers.

Each chapter has several features.

What you need to know

An adaptation from the SQA course specification showing what can be assessed.

Key diagram

A labelled and annotated diagram or image which represents the main ideas in the course specification.

C-type questions

A comprehensive group of short answer questions aimed at giving total coverage of what might be asked at the level of Grade C. These questions typically start with 'Name', 'State', 'Give' or 'Identify'.

A-type questions

A group of longer answer questions which are more difficult and are typical of what might be asked at the level of Grade A. These questions typically start with 'Describe', 'Explain' or 'Suggest'.

Extended response questions

A set of extended response questions for between 4 and 10 marks, containing some straightforward marks linked to grade C and a few more challenging marks linked to grade A.

Exam challenge

 This symbol indicates a topic or idea which has proved challenging for examination candidates in recent years – we give a short hint or comment here.

Cross reference

➡ **This symbol indicates a link with another part of the course or link to model answers and commentary – these are definitely worth following up.**

Technique

 This symbol indicates a named experimental technique with which you need to be familiar for your exam and maybe for your assignment – we give a very brief outline of the technique.

Model answers

SQA-type national standard **model answers** as well as a **commentary** adding more information about the questions and their answers. Note that / and **OR** mean acceptable alternative words and answers respectively.

Glossary flash cards

Flashcards can be downloaded from the Leckie and Leckie website: **https://collins.co.uk/pages/scottish-curriculum-free-resources**. We provide grids with some of the trickier terms, which can be printed, cut and glued to make flash cards.

You can use flash cards to check your knowledge or to test a friend.

How to use this book

We recommend that you use the book all through your course and try the questions in Sections 1–3 as soon as you have completed the topic at school or college. You can also use the book when you are revising for tests and exams. Section 4 could help you in planning your research and the writing of your report. Section 5 can be used to revise individual skills and to test yourself with the example questions provided on-line at **https://collins.co.uk/pages/scottish-curriculum-free-resources**.

Use the grid on pages 4–7 to track your progress. You could attempt all the questions in a chapter, mark them using the Answers section and add your total to the progress grid. This could help you identify areas which require further revision.

Structure and organisation of DNA

What you need to know about DNA structure

DNA is a very long double-stranded molecule in the shape of a double helix.

Each DNA strand is made up from chemical units called **nucleotides**.

A nucleotide is made up of three parts: a deoxyribose sugar, a phosphate and a base.

Deoxyribose molecules have five carbon atoms, which are numbered 1 to 5.

The phosphate of one nucleotide is joined to carbon 5 (5′) of its sugar and linked to carbon 3 (3′) of the sugar of the next nucleotide in the strand to form a 3′–5′ sugar–phosphate backbone.

There are four different bases, called adenine (A), guanine (G), thymine (T) and cytosine (C).

The nucleotides of one strand of DNA are linked to the nucleotides on the second strand through their bases – the bases form pairs joining the strands.

Bases pair in a complementary way – adenine always pairs with thymine and guanine always pairs with cytosine.

Base pairs are held together by hydrogen bonds.

Each strand has a sugar–phosphate backbone with a 3′ end that starts with a deoxyribose molecule and a 5′ end that finishes with a phosphate.

The two strands of a DNA molecule run in opposite directions and are said to be **antiparallel** to each other.

The base sequence of DNA forms the genetic code.

Key diagram

Representation of the main features of a DNA molecule, showing a small piece of DNA with double antiparallel strands joined by complementary base pairs. The length and double helix shape of the molecule are not shown.

Key to bases

A – adenine
T – thymine
G – guanine
C – cytosine

single nucleotide

5′
3′

phosphate

deoxyribose sugar

hydrogen bonds

A T
G C
T A
C G

3′
5′

C-type questions

1. Name the **three** components that make up a single nucleotide.   1
2. Name the **two** components which make up the backbone of each DNA strand.   1
3. Name the complementary base pairs in a DNA molecule.   2
4. Name the type of bond which links the complementary base pairs between DNA strands.   1
5. Give the meaning of the term 'antiparallel' in relation to DNA structure.   1
6. Name the component of the nucleotide located at the 3′ end of each strand.   1
7. Name the component of the nucleotide located at the 5′ end of each strand.   1
8. Name the shape into which a DNA molecule is formed.   1

A-type questions

9. Describe how the two strands of DNA are held together.   2
10. Explain how DNA carries the genetic code.   1

Remember to use the term 'complementary' when describing base pairs – it's the best and most meaningful way of talking about them.

➡ **Model answers and commentary can be found on page 14.**

What you need to know about DNA organisation

Prokaryotes have a single, circular **chromosome** and smaller circular **plasmids**.

Eukaryotes all have linear chromosomes in their nuclei, in which the DNA is tightly coiled and packaged with associated **proteins**.

The associated proteins are called histones.

Eukaryotes also contain circular chromosomes in the **mitochondria** and **chloroplasts** in their cells.

Yeast is a special example of a eukaryote as it also has plasmids.

Key diagram

(a) Prokaryotic cell with a single circular chromosome and plasmids in the cytoplasm. **(b)** Eukaryotic animal cell with linear chromosomes in the nucleus and circular chromosomes in the mitochondrion. **(c)** Eukaryotic plant cell with linear chromosomes in the nucleus and circular chromosomes in the mitochondrion and chloroplast. **(d)** Yeast cell with linear chromosomes in the nucleus, circular chromosome in the mitochondrion and plasmids in the cytoplasm – a special eukaryote.

(a)

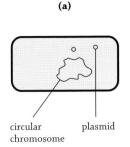

circular   plasmid
chromosome

(b)

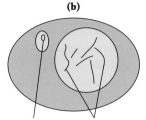

circular   linear chromosomes
chromosome in   associated with histones
mitochondrion   in nucleus

(c)

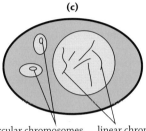

circular chromosomes   linear chromosomes
in mitochondrion and   associated with histones
chloroplast   in nucleus

(d)

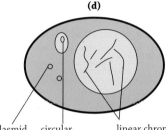

plasmid   circular   linear chromosomes
          chromosome in   associated with histones
          mitochondrion   in nucleus

C-type questions

11. Name **two** structures present in prokaryotic cells that are composed of DNA.   2
12. Name **two** organelles present in a green plant cell that contain circular DNA.   2
13. Give **two** locations, other than the nucleus, where DNA is found in yeast cells.   2
14. Name the molecules associated with the tightly coiled DNA in a eukaryotic cell nucleus.   1
15. Describe why yeast can be thought of as a special example of a eukaryote.   1

A-type questions

16. Describe how DNA is organised within photosynthetic plant cells.   3
17. DNA holds the genetic information in chromosomes in both prokaryotic and eukaryotic cells.

    Describe **two** organisational differences between prokaryotic and eukaryotic chromosomal DNA.   2
18. Cells can be classified as prokaryotic or eukaryotic.

    Describe the organisation and distribution of DNA in the following cell types:

    (a) a prokaryotic bacterium   2

    (b) a eukaryotic plant cell   2

 Candidates find describing how DNA is organised in eukaryotic photosynthetic plant cells very challenging – remember to mention the circular chromosomes in the mitochondria and chloroplasts.

Remember that yeast is a special example of a eukaryotic cell because it also has plasmids – plasmids are found in prokaryotes but also in some special eukaryotes.

Extended response questions

19. Give an account of the structure of a DNA molecule.   5
20. Describe the differences between prokaryotic and eukaryotic cells.   5
21. Describe the function of DNA and give an account of the structure of a
      DNA molecule.   7

Model answers and commentary

Question		Model answer	Marks	Commentary with hints and tips
1		Deoxyribose **AND** phosphate **AND** base	1	Remember to **name** 'deoxyribose' – not just 'sugar'.
2		Deoxyribose and phosphate	1	Again 'deoxyribose' must be stated and **NOT** just 'sugar–phosphate' backbone.
3		Adenine pairs with thymine; guanine pairs with cytosine	2	Remember: **A**pples in **T**rees; **G**arages for **C**ars.
4		Hydrogen	1	Hydrogen bonds link complementary base pairs, and remember – they are also involved in maintaining the 3D shape of proteins.
5		One strand runs from 3′ to 5′ and its complementary strand runs from 5′ to 3′	1	The strands in DNA run in opposite directions – you must mention the 3′ to 5′ and 5′ to 3′. A small diagram might be useful here.
6		Deoxyribose	1	Try drawing out a DNA nucleotide and labelling its 3′ and 5′ ends.
7		Phosphate	1	Phosphate at the 5′ end – think of the phosphate giving a high five!
8		Double helix	1	Easy one but 'double' is an essential part of the answer.
9		Hydrogen bonding; between complementary bases	2	Remember to use the term 'complementary' **OR** 'specific' when referring to the base pairs.

10		The base sequence of DNA forms the genetic code	1	It is the **sequence** or order of bases along one of the strands which forms the code.
11		Circular chromosome; plasmid(s)	2	Make sure you know the organisation of DNA in plants, animals, yeast and prokaryotes.
12		Nucleus; mitochondria; chloroplast **(Any 2)**	2	Candidates often forget about the circular chromosomes in the mitochondria and chloroplasts.
13		Mitochondria; plasmid(s)	2	Yeast is a eukaryote so candidates often forget that yeast, unlike other eukaryotes, may contain plasmids as well.
14		Histones	1	As soon as you read 'DNA associated with', think histones! Saying 'proteins' is not strong enough.
15		It has plasmids	1	Yeast cells are eukaryotic but have plasmids, which are more usually features of prokaryotic cells.
16		Linear chromosomes in nucleus; circular chromosomes in mitochondria; circular chromosomes in chloroplasts	3	The question asks about photosynthetic plant cells and is prompting you to refer to the chloroplasts as well.
17		Linear chromosomes in eukaryotes and circular chromosome in prokaryotes **OR** eukaryotic chromosomes located in the nucleus but prokaryotes have no nucleus **OR** eukaryotic chromosomes are associated/tightly coiled with protein/histones **OR** eukaryotes contain circular chromosomes in mitochondria/chloroplasts **(Any 2)**	2	Try writing this out a few times or arranging as a table.
18	(a)	Prokaryotes have circular chromosomes; and plasmids	2	Remember that prokaryotes do not contain a nucleus.
	(b)	Eukaryotic DNA is packaged in linear chromosomes in the nucleus; and in circular chromosomes in chloroplasts/ mitochondria; linear chromosomes are associated/ tightly coiled with protein **(Any 2)**	2	Lots of variations of DNA organisation questions are possible, so it is worth becoming an expert!

19	1	Double helix	5
	2	Chains/strands composed of nucleotides	
	3	Nucleotide is deoxyribose, a phosphate and a base	
	4	Sugar–phosphate backbone	
	5	Complementary base pairing **OR** A with T, G with C	
	6	Antiparallel chains/strands **OR** each chain has a deoxyribose at its 3′ end and a phosphate at its 5′ end **(Any 5)**	
20	1	Circular chromosomal DNA in prokaryotic cells	5
	2	Plasmids in prokaryotic cells	
	3	Plasmids in yeast	
	4	Circular chromosome in mitochondria of eukaryotic cells	
	5	Circular chromosome in chloroplasts of eukaryotic cells	
	6	DNA in the linear chromosomes of the nucleus of eukaryotic cells	
	7	DNA is tightly coiled and packaged with associated proteins in eukaryotic cells **(Any 5)**	
21		Function:	7
	1	Carries the genetic code	
	2	In the base sequence	
	3	The order of bases along one strand **(Any 2)**	
		Structure:	
	4	Double helix	
	5	Chains/strands of nucleotides	
	6	Nucleotide is deoxyribose, a phosphate and a base	
	7	Sugar–phosphate backbones	
	8	Complementary base pairing **OR** A with T, G with C	
	9	Antiparallel chains/strands **(Any 5)**	

Replication of DNA

What you need to know about DNA replication

Prior to cell division, DNA is replicated by a **DNA polymerase**.

DNA polymerase needs primers to start replication.

A primer is a short strand of nucleotides which binds to the 3′ end of the template DNA strand allowing polymerase to add DNA nucleotides.

DNA polymerase adds DNA nucleotides, using complementary base pairing, to the deoxyribose (3′) end of the new DNA strand which is forming.

To start replication, DNA is unwound and hydrogen bonds between bases are broken to form two template strands.

DNA polymerase can only add DNA nucleotides in one direction, resulting in the **leading strand** being replicated continuously and the **lagging strand** being replicated in fragments.

Fragments of DNA are joined together by **ligase**.

Key diagram

Labelled representation of DNA replication showing two template strands, primers and the action of DNA polymerase and DNA ligase. The **3′–5′** leading strand is replicated continuously and the lagging strand is replicated in fragments as shown.

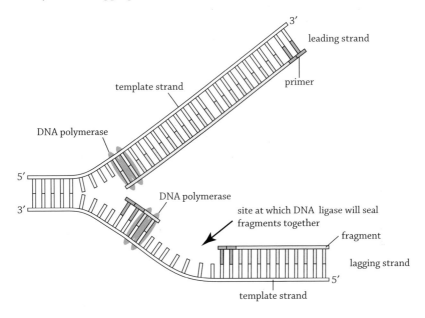

C-type questions

1. Other than the original DNA template and ATP, name **two** requirements for DNA replication.   2

2. Give the role of DNA polymerase in the process of DNA replication.   1

3. Name the short sequences of bases needed to start DNA replication.   1

4. Describe the end of the DNA strand to which DNA polymerase adds DNA nucleotides.   1

5. Name the enzyme responsible for joining DNA fragments on the lagging strand together during replication.   1

6. Name the **two** enzymes required in the replication of the lagging strand of a DNA molecule.   2

7. Describe the direction of replication on each of the template strands of DNA.   1

A-type questions

8. During DNA replication two new daughter strands are synthesised using the original strands as templates.

   Explain why the antiparallel nature of the DNA molecule results in the complementary copy of the lagging strand being synthesised in short fragments.   2

> **!** Many candidates find it tricky to explain why the antiparallel nature of DNA affects how DNA polymerase synthesises a copy of its template strands – remember that the direction of replication of DNA is from 3′ to 5′ on both template strands.

9. Explain why cells need to carry out DNA replication.   2

> **!** Many candidates find difficulty in explaining why DNA must replicate – remember that DNA replication occurs prior to cell division and the daughter cells produced each need copies of the parental DNA.

➡ **Model answers and commentary can be found on page 20.**

What you need to know about polymerase chain reaction

Polymerase chain reaction (PCR) amplifies DNA using complementary **primers** for specific target sequences.

In PCR, primers are short strands of nucleotides which are complementary to specific target sequences at the two ends of the region of DNA to be amplified.

Repeated cycles of heating and cooling amplify the target region of DNA.

DNA is heated to between 92 and 98°C to separate the strands.

It is then cooled to between 50 and 65°C to allow primers to bind to target sequences.

It is then heated to between 70 and 80°C for **heat-tolerant DNA polymerase** to replicate the region of DNA.

Practical applications of PCR include the amplification of DNA to help solve crimes, settle paternity suits and diagnose genetic disorders.

Key diagram

Graph of one thermal cycle in PCR – Stage X: DNA strands separate; Stage Y: primers bind; Stage Z: heat-tolerant DNA polymerase activity

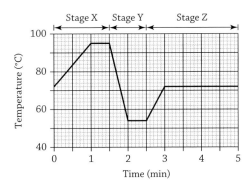

Technique

Using **gel electrophoresis** to separate macromolecules such as DNA fragments is a technique you need to familiar with for your exam.

Once DNA has been amplified by PCR it can be broken down into fragments and stained. The stained fragments can be separated by gel electrophoresis to produce a unique DNA profile for an individual.

C-type questions

10. State the purpose of PCR.   1

11. Name the short sequences of bases needed to start DNA replication.   1

12. Name the enzyme which replicates the section of DNA to be amplified in PCR.   1

13. Give **one** practical application of PCR.   1

A-type questions

14. PCR amplifies specific sequences of DNA.

    Explain the purpose of the different heat treatments in a cycle
    of the PCR procedure.   3

15. The flow chart below shows temperature changes during steps in the PCR procedure.

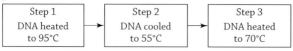

(a) Describe the effect of the increase in temperature at step 1 on the structure of DNA. 1

(b) Explain why the DNA polymerase used in step 3 can function at 70°C although the high temperature would denature most enzymes. 1

16. Describe how the strands of DNA are separated during PCR. 1

17. Explain why primers are required in PCR. 1

18. Describe the role of primers in DNA replication during PCR and explain why two different primers are needed. 2

Extended response questions

19. Describe the role of enzymes in DNA replication. 4

20. Describe the main steps in PCR. 4

21. Give an account of the replication of a molecule of DNA. 7

Model answers and commentary

Question		Model answer	Marks	Commentary with hints and tips
1		Supply of DNA nucleotides; primer; DNA polymerase; DNA ligase **(Any 2)**	2	Remember to specify **DNA** nucleotides.
2		DNA polymerase adds nucleotides to 3'/ deoxyribose end of strand	1	DNA is a polymer made up of DNA nucleotides. Many enzymes end in 'ase', so the enzyme in DNA replication is DNA polymerase.
3		Primer	1	Primer is complementary to a target sequence of bases on the 3' end of the DNA strand.
4		3'/deoxyribose end of strand	1	It is worth noting that DNA polymerase adds DNA nucleotides to the 3' end of the DNA **OR** 3' end of the primer **OR** in a 3' to 5' direction.

5		DNA ligase	1	Remember **LL** = **L**igase on the **L**agging strand. **LIG**ase on the **LAGG**ing strand.
6		DNA polymerase; DNA ligase	2	DNA polymerase to add the DNA nucleotides and ligase to join the fragments.
7		Direction of DNA replication is from 3′ to 5′ on both template strands	1	Continuously on the leading strand and in fragments on the lagging strand.
8		DNA polymerase adds nucleotides to 3′/ deoxyribose end of strand; primers bind to lagging strand in many places	2	Primers attach to 3′ end and so only one is needed on the leading strand but many on the lagging strand as the DNA gradually unzips.
9		Cell division/mitosis; ensures each daughter cell receives all the genetic information needed to carry out all of their functions	2	DNA replication occurs prior to cell division and the daughter cells produced each need copies of the parental DNA.
10		Amplification of DNA/to generate many copies of a DNA sequence	1	Remember the term 'amplification'.
11		Primer	1	A primer is a short strand of RNA or DNA (generally about 18–22 bases long) that serves as a starting point for DNA synthesis. DNA polymerases can only add new nucleotides to an existing strand of DNA. Primer is complementary to a target sequence of bases on the 3′ end of the DNA strand.
12		DNA polymerase **OR** Taq polymerase	1	DNA is a polymer composed of nucleotides. Many enzymes end in 'ase', so the enzyme in DNA replication is DNA polymerase. This is a heat-tolerant version of the enzyme used to replicate DNA naturally in most organisms.

13		Amplified DNA can be used in forensics/solving crimes/paternity tests/maternity tests **OR** can provide sequence data/show place of extinct species in evolution **OR** diagnose genetic disorders	1	These are the uses made of the DNA **after** it has been amplified.
14		The DNA is heated to 92–98°C to denature the DNA/separate the DNA strands; The temperature is cooled/lowered to 50–65°C to allow the primers to bind to the target sequences; The temperature is then raised to 70–80°C for DNA polymerase to synthesise new strands	3	Remember **PCR '967'** – for example: **95**°C to separate DNA strands **65**°C to allow primers to attach **75**°C as optimum temperature for DNA polymerase.
15	(a)	The DNA is heated to 95°C to denature the DNA/separate the DNA strands/break the hydrogen bonds between the complementary base pairs	1	It is good practice to use the term 'complementary' each time you mention base pairs.
	(b)	Heat-tolerant DNA polymerase/Taq polymerase is obtained from bacteria adapted to live in hot springs	1	DNA polymerase from most organisms would be denatured at the high temperatures needed for PCR.
16		The DNA is heated to 90°C/95°C to denature the DNA/separate the DNA strands/break the hydrogen bonds between the complementary base pairs	1	It is good practice to use the term 'complementary' each time you mention base pairs.
17		DNA polymerase requires primers to start replication **OR** DNA polymerase can only add complementary DNA nucleotides to an existing strand **OR** Primers are short complementary sequences of nucleotides that allow binding of DNA polymerase	1	Primers serve as the starting point for DNA synthesis. DNA polymerases can only add new nucleotides to an existing strand of DNA. Primer is complementary to a target sequence of bases on the 3′ end of the DNA strand.

18		DNA polymerase requires primers to start replication **OR** Primers are short complementary sequences of nucleotides that allow binding of DNA polymerase; Two different primers are needed to match the two different complementary target sequences at the 3′ ends of the DNA strands	2	Candidates often miss out on achieving the second mark, which requires them to give the reason for the need for two different primers in the PCR process.
19	1	DNA polymerase	4	
	2	Adds DNA nucleotides to 3′/deoxyribose end of strand		
	3	DNA polymerase adds DNA nucleotides in one direction		
	4	DNA ligase		
	5	Joins fragments on the lagging strand		
	6	Leading strand replicated continuously, the lagging strand in fragments **(Any 4)**		
20	1	DNA heated to 90°C to denature the DNA/separate the DNA strands/break the hydrogen bonds between the complementary base pairs	4	
	2	Cooled to 60°C to allow primers to bind to target sequences		
	3	Primers are complementary to specific target sequences at the two ends of the region to be amplified/3′ ends of the DNA strands		
	4	Heat-tolerant DNA polymerase then replicates the primed region of DNA at 70°C		
	5	Repeated cycles of heating and cooling amplify this region of DNA **(Any 4)**		
21	1	DNA uncoils and unzips	7	
	2	Primers bind at end of leading template strand/3′ end of the DNA strand		
	3	DNA polymerase adds complementary DNA nucleotides to leading strand continuously/3′ end of the primer molecule		
	4	Primers bind to lagging strand in many places		
	5	DNA polymerase adds complementary DNA nucleotides to lagging strand in fragments		
	6	Fragments joined by DNA ligase		
	7	Replication occurs at several positions on a DNA molecule at the same time		
	8	Replication requires energy/ATP **(Any 7)**		

CHAPTER 3

Control of gene expression

What you need to know about RNA

Only a fraction of the genes in any cell are expressed.

Gene expression involves the **transcription** and **translation** of DNA sequences.

Transcription and translation involve three types of RNA (mRNA, tRNA and rRNA).

RNA is single stranded and is composed of nucleotides containing ribose sugar, phosphate and one of four bases: cytosine, guanine, adenine and uracil.

Messenger RNA (mRNA) carries a copy of the DNA code from the nucleus to the ribosome.

mRNA is transcribed from DNA in the nucleus and translated into proteins by ribosomes in the cytoplasm.

Each triplet of bases on the mRNA molecule is called a **codon** and codes for a specific amino acid.

Transfer RNA (tRNA) folds due to complementary base pairing.

Each tRNA molecule carries its specific amino acid to the ribosome.

Ribosomal RNA (rRNA) and proteins form ribosomes.

A tRNA molecule has an **anticodon** (an exposed triplet of bases) at one end and an attachment site for a specific amino acid at the other end.

Key diagram

Summary of gene expression – transcription of DNA and translation into protein

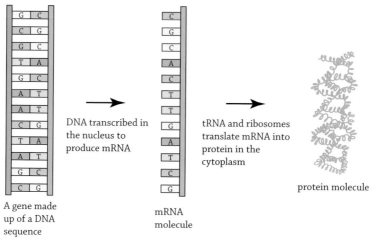

C-type questions

1. Name **two** processes which are involved in gene expression.   2
2. Name the molecule produced by the transcription of DNA.   1
3. Name the molecules produced by the translation of mRNA by ribosomes and tRNA.   1

A-type question

4. Describe the locations of transcription and translation in gene expression.   2

➡ **Model answers and commentary can be found on page 30.**

Key diagram

The three types of RNA and their functions – mRNA, tRNA and a ribosome made up of rRNA and proteins

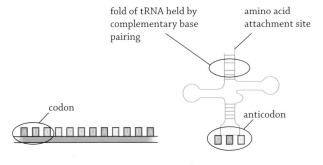

fold of tRNA held by complementary base pairing

amino acid attachment site

codon

anticodon

mRNA carries a complementary copy of the sequence of bases on DNA to the ribosome

tRNA carries its specific amino acid to the ribosome

rRNA and proteins form the ribosome which is the site of translation

C-type questions

5. Give **two** differences between the structure of RNA and that of DNA.   2
6. Name the molecule which carries a complementary copy of DNA from the nucleus to the ribosome.   1
7. Name the molecule which carries specific amino acids to the ribosome.   1
8. Name **one** substance, other than rRNA, that makes up the ribosome.   1

A-type questions

9. Describe the function of mRNA.   2
10. Describe the function of tRNA.   2
11. Describe the general structure of a molecule of tRNA.   3

Many candidates find describing the structure of tRNA difficult – remember to include anticodon, amino acid attachment site and folding due to complementary base pairing.

➡ **Model answers and commentary can be found on page 30.**

What you need to know about transcription and splicing

RNA polymerase moves along DNA unwinding the double helix and breaking the hydrogen bonds between the bases.

RNA polymerase synthesises a **primary transcript** of mRNA from RNA nucleotides using complementary base pairing.

Uracil in RNA is complementary to adenine.

RNA splicing forms a mature mRNA transcript.

The **introns** of the primary transcript are non-coding regions and are removed.

The **exons** are coding regions and are joined together to form the mature transcript.

The order of the exons is unchanged during splicing.

Different proteins can be expressed from one gene as a result of alternative RNA splicing.

Different mature mRNA transcripts are produced from the same primary transcript depending on which exons are retained.

Key diagram

Summary of transcription, showing unwinding, unzipping and addition of free nucleotides to the template stand by RNA polymerase

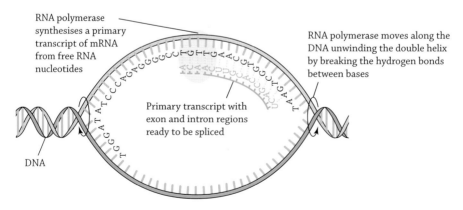

RNA polymerase synthesises a primary transcript of mRNA from free RNA nucleotides

RNA polymerase moves along the DNA unwinding the double helix by breaking the hydrogen bonds between bases

Primary transcript with exon and intron regions ready to be spliced

DNA

C-type questions

12. Give the location of mRNA synthesis in a cell. 1
13. Give the term used to describe the production of a primary transcript from a DNA template. 1
14. Name the enzyme required for the synthesis of a primary transcript from RNA nucleotides during transcription. 1

A-type question

15. Describe the function of RNA polymerase in the synthesis of a primary transcript. 3

Many candidates fail to fully describe the function of RNA polymerase – remember it unwinds and breaks hydrogen bonds between DNA bases as well as adding complementary nucleotides to the primary transcript.

➡ **Model answers and commentary can be found on page 31.**

Key diagram

Alternative splicing of a primary transcript

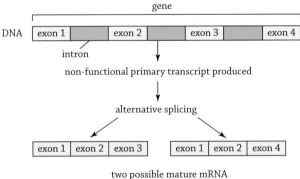

Note that the mature mRNA transcripts depend on which exons have been retained and that their order is unchanged

C-type questions

16. Name the sequences of bases in eukaryotic genes which **do not** code for proteins. 1
17. Name the sequences of bases in eukaryotic genes which **do** code for proteins. 1
18. State the difference between introns and exons. 1
19. Name the non-functional mRNA molecule which contains both coding and non-coding regions. 1
20. Name the process that involves the modification of the primary transcript to form the functional mRNA. 1

A-type questions

21. In eukaryotic cells, mRNA is spliced after transcription.

    Describe what happens during RNA splicing.   3

22. In some eukaryotic cells, different mRNA molecules, and therefore different proteins, can be expressed from a single gene.

    Name and describe the process which results in different mRNA molecules being expressed.   2

23. Explain the significance of alternative RNA splicing in terms of gene expression.   2

 Many candidates find alternative RNA splicing difficult – remember that different mature mRNA transcripts can be produced from a primary transcript depending on which exons are retained.

➡ **Model answers and commentary can be found on page 31.**

What you need to know about translation

tRNA is involved in the translation of mRNA into a **polypeptide** at a ribosome.

Translation begins at a start codon and ends at a stop codon.

Anticodons bond to codons by complementary base pairing, translating the genetic code into a sequence of amino acids.

Peptide bonds join the amino acids together.

Each tRNA then leaves the ribosome as the polypeptide is formed.

Key diagram

Summary of translation to synthesise a polypeptide, showing tRNA aligning specific amino acids using codon–anticodon matching at a ribosome. Note the presence of start and stop codons.

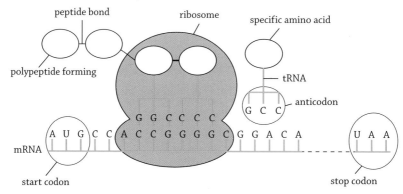

C-type questions

24. Give the location of mRNA translation in a cell.   1
25. State the number of bases in an mRNA molecule which code for an individual
    amino acid in a polypeptide.   1
26. Name the triplet of bases on an mRNA molecule that codes for an amino acid.   1
27. Name the triplet of bases on a tRNA molecule that codes for a specific amino acid. 1
28. State what causes the translation of an mRNA molecule to terminate.   1

A-type questions

29. Describe how the structure of tRNA is related to its function.   2
30. Describe the role of tRNA in translation of mRNA.   1
31. Explain the importance of the tRNA anticodon in the process of translation.   2

➡ **Model answers and commentary can be found on page 32.**

What you need to know about protein structure

Amino acids are linked by peptide bonds to form polypeptides.

Polypeptide chains fold to form the three-dimensional shape of a protein, held together by hydrogen bonds and other interactions between individual amino acids.

Proteins have a large variety of shapes which determine their functions.

Phenotype is determined by the proteins produced as the result of gene expression.

Environmental factors also influence phenotype.

Key diagram

Protein made up of a folded polypeptide chain held in a three-dimensional shape by hydrogen bonds or other interactions between individual amino acids

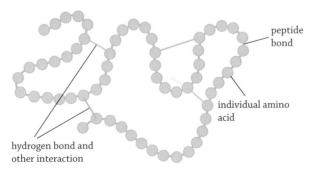

peptide bond

individual amino acid

hydrogen bond and other interaction

C-type questions

32. Name the bonds which connect the amino acids to form a polypeptide chain.   1
33. State how protein molecules are held in three-dimensional shapes.   1
34. State how phenotype is determined in terms of gene expression.   1
35. Apart from gene expression, give **one** other factor which can influence phenotype   1

A-type question

36. Explain the importance of the three-dimensional shape of a protein molecule.   1

Extended response questions

37. Give an account of how different proteins can be produced from the expression of the same gene.   4
38. Give an account of protein structure.   4
39. Give an account of gene expression under the following headings:

    (a) formation of mature mRNA   5
    (b) translation of mature mRNA   4

Model answers and commentary

Question		Answer	Marks	Commentary with hints and tips
1		Transcription; translation	2	Transcription is the copying of a DNA sequence to make a primary transcript. Translation is the production of a polypeptide using sequences of mRNA.
2		Messenger RNA (mRNA)	1	The abbreviation is OK here.
3		Proteins	1	Only a fraction of the genes in an individual cell are expressed.
4		Transcription occurs in the nucleus of cells; translation occurs on ribosomes in the cytoplasm	2	The transcription and translation processes involve the three type of RNA (mRNA, tRNA and rRNA).

5		RNA is single stranded **AND** DNA is double stranded **OR** RNA has ribose **AND** DNA has deoxyribose **OR** RNA has uracil **AND** DNA has thymine **(Any 2)**	2	RNA molecules are differently shaped to DNA molecules and their nucleotides have different sugars: in RNA the base uracil replaces thymine as complementary to adenine. Remember, the question referred to structural differences – you could be asked about functional differences too so make sure you know them!
6		Messenger RNA (mRNA)	1	Remember that the mRNA is a complementary copy of the gene code and carries it to the ribosomes to be translated.
7		Transfer RNA (tRNA)	1	Remember that each tRNA molecule carries its **specific** amino acid.
8		Protein	1	Ribosomes are composed of rRNA and protein.
9		Carries a complementary copy of the sequence of bases of DNA; from the nucleus to the ribosome	2	The sequence of DNA bases in a gene determines the order of amino acids in the protein synthesised.
10		Transfers specific amino acids from the cytoplasm; to the mRNA on the ribosome	2	Must be clear that tRNA carries its own **specific** amino acid.
11		Single folded strand of RNA nucleotides held by complementary bases pairs; triplet anticodon site; amino acid attachment site	3	Practise drawing and labelling a diagram of tRNA.
12		Nucleus	1	DNA is inside the nucleus and so transcription of a gene must take place in the nucleus.
13		Transcription	1	A transcript is a copy. The copying process is called transcription.
14		RNA polymerase	1	The primary transcript is a polymer made up of RNA nucleotides, so the enzyme that synthesises it is called RNA polymerase.

15		Moves along DNA unwinding the double helix; breaking hydrogen bonds between bases; adds free complementary RNA nucleotides to the primary transcript	3	Remember to give all three actions of RNA polymerase.
16		Introns	1	Remember – **ex**ons are **ex**pressed and **in**trons **in**terrupt the gene.
17		Exons	1	Remember – **ex**ons are **ex**pressed and **in**trons **in**terrupt the gene.
18		Exons are coding regions and introns are non-coding regions.	1	Remember – coding means coding for protein.
19		Primary transcript	1	The primary transcript is the first copy/version of the mRNA and needs to have the introns removed to be the mature transcript.
20		RNA splicing	1	Modification = change. To become functional the introns need to be removed from the primary transcript.
21		Introns are removed from the primary transcript; exons are joined/spliced together to produce the mature mRNA; the order of exons is unchanged during splicing	3	It is important to remember to mention that the order of exons is unchanged.
22		Alternative RNA splicing; depends on which exons are retained/not all exons are used	2	'Alternative' RNA splicing rather than just RNA splicing.
23		Allows different mature transcripts; so that different proteins can be produced from a single gene/DNA sequence	2	Significance means **importance**.
24		Ribosomes	1	Slightly more demanding because it requires you to know the term 'mRNA translation'.
25		3	1	Triplet code.

26		Codon	1	Each codon codes for a specific amino acid.
27		Anticodon	1	A sequence of three bases on the tRNA molecule complementary to a codon on the mature mRNA.
28		Stop codon	1	Codon which causes translation to finish or terminate when the polypeptide is complete.
29		It has an amino acid attachment site that binds to/transports a specific amino acid; It has a triplet of bases called an anticodon that are complementary/match to an mRNA codon	2	Note the importance of including the terms 'specific' and 'complementary'.
30		The tRNA carries/transfers specific amino acids to the mRNA on the ribosome	1	Once again the term 'specific' is needed.
31		The tRNA anticodons are complementary to the codons of the mRNA; this ensures that the amino acids are joined in the correct order to synthesise the protein	2	'Explain' question needs more than just the role of the tRNA.
32		Peptide	1	Poly**peptide**s have **peptide** bonds.
33		Hydrogen bonds **OR** other interactions	1	Hydrogen bond may be the easiest to remember as you have come across this bond already in complementary base pairing.
34		By the proteins produced	1	Remember that genes are expressed to produce proteins and that proteins help determine the phenotype of the organism.
35		Environmental factors	1	Remember that environmental factors such as nutrition, light and temperature can influence the final phenotype.
36		Protein function depends on the three-dimensional shape of its molecules	1	Think proteins, think shape! Remember how shape is important in the active site of an enzyme, receptor proteins and hormones.

37		1	Alternative RNA splicing	4
		2	Different mature RNA transcription produced from the same primary transcript	
		3	Depending on which exons are retained	
		4	Different base sequences in mature transcripts	
		5	Different amino acid sequences in polypeptide produced **(Any 4)**	
38		1	Made of a chain of amino acids	4
		2	Held by peptide bonds	
		3	Polypeptide folded	
		4	To form three-dimensional shape	
		5	Held together by hydrogen bonds and other interactions between individual amino acids **(Any 4)**	
39	(a)	1	RNA polymerase unwinds the double helix/DNA/gene	5
		2	Hydrogen bonds broken between bases	
		3	RNA polymerase adds complementary RNA nucleotides	
		4	(To synthesise a) primary transcript of mRNA	
		5	Introns removed	
		6	Exons spliced to make mature mRNA **(Any 5)**	
	(b)	1	Mature mRNA goes to ribosome	4
		2	tRNA carries their specific amino acids	
		3	Anticodons on tRNA aligned with codons on mRNA	
		4	Amino acids aligned in correct sequence	
		5	Amino acids linked by peptide bonds **(Any 4)**	

4 Cellular differentiation

What you need to know about differentiation and where it happens

Cellular **differentiation** is the process by which a cell expresses certain genes to produce proteins characteristic for that type of cell.

Differentiation allows a cell to carry out specialised functions.

Differentiation into specialised cells occurs in meristems in plants.

Meristems are regions of unspecialised cells in plants that can divide (self-renew) and/or differentiate.

Differentiation into specialised cells occurs in **embryonic** and **tissue (adult) stem cells** in animals.

Stem cells are unspecialised cells in animals that can divide (self-renew) and/or differentiate.

Key diagram

Stem cell dividing to produce more stem cells (self-renewal) and differentiating into specialised cells

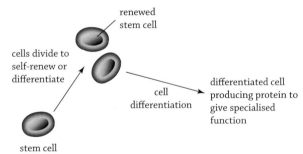

renewed
stem cell

cells divide to
self-renew or
differentiate

cell
differentiation

differentiated cell
producing protein to
give specialised
function

stem cell

C-type questions

1. Give the term used to describe the process by which a cell develops specialised functions.   1
2. Name the regions of unspecialised cells in plants capable of cell division.   1

A-type questions

3. Describe what is meant by the term 'differentiation'.   2
4. After a stem cell differentiates, only certain genes are expressed.
   Explain how this results in different cell types.   1

Many candidates struggle to explain how a cell differentiates into a specialised cell – remember to say that it expresses certain genes to produce proteins characteristic of that cell type.

➡ **Model answers and commentary can be found on page 38.**

What you need to know about stem cells and their functions and uses in medicine

Cells in the very early embryo can differentiate into all the cell types that make up the organism and so are **pluripotent**.

All the genes in embryonic stem cells can be switched on so these cells can differentiate into any type of cell.

Tissue stem cells are involved in the growth, repair and renewal of the cells found in that tissue.

Tissue stem cells are **multipotent** as they can differentiate into all of the types of cell found in a particular tissue type. For example, blood stem cells located in bone marrow can give rise to all types of blood cell.

Stem cells have both **therapeutic** and **research uses**.

Therapeutic uses of stem cells involve the repair of damaged or diseased organs or tissues.

Stem cells are used in corneal repair and the regeneration of damaged skin.

Stem cells from the embryo can self-renew, under the right conditions, in the laboratory.

Research uses of stem cells involve them being used as model cells to study how diseases develop or being used for drug testing.

Stem cell research provides information on how cell processes such as cell growth, differentiation and gene regulation work.

The use of embryonic stem cells can offer effective treatments for disease and injury; however, it raises ethical issues because it involves the destruction of embryos.

Key diagram

Properties of different types of stem cell

Type of stem cell	Level of potency	Description
Embryonic	Pluripotent	Can differentiate into any type of cell during development
Tissue (adult)	Multipotent	Can differentiate into any types of cell found in a particular tissue type during growth, repair and renewal

C-type questions

5. Give the characteristic which is unique to embryonic stem cells.   1
6. Name the type of stem cell capable of differentiating into all the types
    of cell that make up the organism to which it belongs.   1

A-type questions

7. Describe the importance of adult stem cells to the human body.   1
8. Describe how tissue (adult) stem cells differ from embryonic stem cells.   2

➡ Model answers and commentary can be found on page 38.

Key diagram

Different uses of stem cells

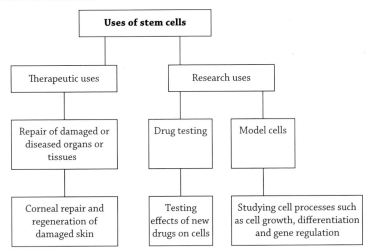

C-type questions

9. Give **one** use of stem cells in medical research.   1
10. Give **one** therapeutic use of stem cells.   1

A-type questions

11. Describe the difference between therapeutic and research uses for stem cells.   2

> **!** Many candidates struggle to explain the difference between
> therapeutic and research uses of stems cells – remember to link
> therapy with treatment of patients.

12. Explain **one** ethical dilemma relating to the use of embryonic stem cells.   2

Extended response questions

13. Describe the differences between and similarities of embryonic stem cells and tissue stem cells.   4

14. Give an account of stem cells under the following headings:

    (a) types of stem cell and their properties   5

    (b) therapeutic use of stem cells and ethical issues surrounding their use   3

Model answers and commentary

Question		Answer	Marks	Commentary with hints and tips
1		Differentiation	1	When different genes have been switched on to allow the cell to produce proteins characteristic of that cell type.
2		Meristems	1	The meristems of a flowering plant include those found at the root and shoot tips.
3		The process by which a cell develops more specialised functions; by expressing certain genes/producing proteins characteristic of that type of cell	2	When asked about differentiation you should mention the term 'specialisation' and vice versa. Make sure that you learn the second part of this answer.
4		The genes that are expressed code for/produce proteins which are characteristic of that type of cell	1	This is a very tricky question and many students do not have the correct phrase to use. It is well worth learning if you are going to boost your grade.
5		They can divide (self-renew) **AND** differentiate into all the cell types that make up the organism **OR** they are pluripotent	1	Remember '**DD**' for stem cells' features: they can **D**ivide and **D**ifferentiate.

6		Embryonic stem cells	1	Tissue (adult) stem cells can only differentiate to form cells characteristic of the tissue in which they originated/give rise to more limited cell types/are multipotent.
7		They can divide and differentiate to replace damaged cells of the type in which they are found **OR** to form cells characteristic of the tissue in which they originated **OR** give rise to more limited cell types	1	You need to be aware that 'tissue' stem cells are also called 'adult' stem cells.
8		Embryonic stem cells can differentiate into almost all cell types/pluripotent; Tissue (adult) stem cells can only differentiate to form cells characteristic of the tissue in which they originated/give rise to more limited cell types/multipotent	2	You need to be able to compare both types of stem cell.
9		To provide information on cell processes/cell growth/differentiation/ gene regulation **OR** used as model cells to study how diseases develop **OR** used as model cells for drug testing	1	You need to learn a therapeutic use and a research use of stem cells. Take care not to mix them up.
10		Replacing/repairing damaged or diseased tissues/organs **OR** Example: skin graft, corneal graft, bone marrow transplant	1	You need to learn a therapeutic use and a research use of stem cells. Take care not to mix them up.
11		Therapeutic use includes the repair of damaged/diseased organs/tissue/cells; Research use includes using stem cells to provide information on cell processes/cell growth/differentiation/ gene regulation **OR** used as model cells to study how diseases develop **OR** used as model cells for drug testing	2	Try to associate certain words and phrases. Link therapeutic with treatment **AND** link research with model cells and drug testing.
12		Embryonic stems cells can offer effective treatment for disease/injury; but this involves the destruction of embryos	2	This model answer is a useful phrase to help you answer this type of question.

13			Similarities:	4
	1		Unspecialised cells	
	2		Can divide and differentiate	
			Differences:	
	3		Embryonic stem cells found in developing embryo	
	4		Embryonic stem cells have the capacity to become all cell types/pluripotent.	
	5		Tissue stem cell are found in body tissues	
	6		Tissue stem cells can only differentiate into cell types from the tissue in which they are found/multipotent **(Any 4)**	
14	(a)	1	Stems cells are unspecialised/undifferentiated cells (in animals)	5
		2	Stem cells divide/self-renew to form new stem cells	
		3	They differentiate/develop into specialised cells/different cells for different functions	
		4	There are tissue (adult) and embryonic stem cells	
		5	Embryonic stem cells can differentiate into almost all cell types/pluripotent	
		6	Tissue (adult) stem cells can differentiate to form cells characteristic of the tissue in which they originated **(Any 5)**	
	(b)	1	Can be used to replace damaged/diseased cells/organs/tissues	3
		2	E.g. skin graft, corneal graft, bone marrow transplant	
		3	Stem cell treatment allows doctors to alleviate suffering	
		4	Obtaining embryonic stems cells involves destruction of embryos **(Any 3)**	

The structure of the genome

What you need to know about the genome

The **genome** of an organism is its entire hereditary information encoded in DNA.

DNA sequences that code for protein are defined as **genes**.

A genome is made up of genes and other DNA sequences that do not code for proteins.

Most of the eukaryotic genome consists of **non-coding sequences**.

Non-coding sequences regulate transcription and others are transcribed but never translated.

tRNA and rRNA are non-translated forms of RNA.

Key diagram

Summary of the structure of the genome, which is the entire hereditary information encoded in an organism's DNA

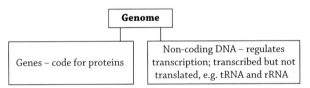

C-type questions

1. Give the term used to describe the entire hereditary information encoded in the DNA of an organism.   1

2. Name the substances which are coded for by genes.   1

3. Give **one** function of non-coding sequences in DNA.   1

4. Name **one** form of RNA which is not translated.   1

A-type question

5. Explain the differences between coding and non-coding DNA sequences.   2

Extended response question

6. Describe the structure of the genome.   5

Model answers and commentary

Question		Answer	Marks	Commentary with hints and tips
1		Genome	1	A genome contains genes that code for proteins and other DNA sequences that do not code for proteins.
2		Proteins	1	Genes are the coding regions of the genome. Coding sequences in the genome are transcribed and translated so mRNA is needed.
3		Regulation of transcription/ regulatory sequences/turning genes on or off **OR** transcribed but not translated/transcribed into tRNA/ transcribed into rRNA	1	Non-coding sequences include those that regulate transcription and those that are transcribed into mRNA but are not translated.
4		tRNA **OR** rRNA	1	Two good examples of the role of the non-coding regions of the genome.
5		Coding sequences/genes code for proteins/amino acid sequences in proteins; Non-coding sequences are regulatory sequences/regulate transcription/ turn genes on or off **OR** Non-coding sequences are transcribed but not translated/transcribed into tRNA/ transcribed into rRNA	2	The function of the non-coding region of the genome is often poorly answered by candidates.
6	1	The genome of an organism is its entire hereditary information encoded in DNA	5	
	2	There are coding and non-coding sequences		
	3	DNA sequences that code for proteins are called genes		
	4	Non-coding sequences include those that regulate transcription		
	5	And those that are transcribed to RNA but are not translated		
	6	Non-translated forms of RNA include tRNA and rRNA **(Any 5)**		

6 Mutation

What you need to know about the effects of mutation

Mutations are changes in the DNA that can result in no protein or an altered protein being synthesised.

C-type question

1. Give the name for a random change in the genome that can result in an altered protein being synthesised. 1

A-type question

2. Explain what is meant by a mutation. 1

➡ **Model answers and commentary can be found on page 47.**

What you need to know about gene mutations

Single gene mutations involve the alteration of a DNA nucleotide sequence as a result of the **substitution**, **insertion** or **deletion of nucleotides**.

Nucleotide substitutions include **missense**, **nonsense** and **splice-site mutations**.

Missense mutations result in one amino acid being changed for another, which may result in a non-functional protein or have little effect on the protein.

Nonsense mutations result in a premature stop codon being produced, which results in a shorter protein.

Splice-site mutations result in some introns being retained and/or some exons not being included in the mature transcript.

Nucleotide insertions or deletions result in frameshift mutations.

Frameshift mutations cause all of the codons and all of the amino acids after the mutation to be changed.

Frameshift mutations have a major effect on the structure of the protein produced.

Key diagram

Mutations which affect single genes – single nucleotide substitution, insertion and deletion

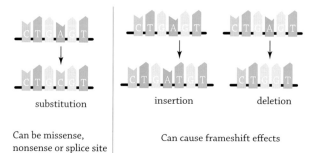

substitution	insertion deletion
Can be missense, nonsense or splice site	Can cause frameshift effects

C-type questions

3. State what is meant by a single gene mutation.   1

> **Many candidates find defining a single gene mutation difficult – remember that they all involve changes in the nucleotide sequence.**

4. Name **two** single gene mutations.   2
5. Give the meaning of the following types of mutation:
    (a) nonsense mutation   1
    (b) splice-site mutation   1

A-type questions

6. Explain the effect of a substitution mutation on the structure of the protein produced.   2
7. Explain how a frameshift mutation affects a gene.   2
8. Describe the effects of a deletion mutation on the structure of the polypeptide coded for by the original gene.   2
9. Describe the effects of an insertion mutation on the structure of the polypeptide coded for by the original gene.   2

> **Many candidates find describing the effects of a deletion tricky – a good word to include in any answer is 'frameshift'.**

10. Describe the effects of a splice-site mutation.   2

11. Describe the effects of a missense mutation.   2

12. Describe the effects of a nonsense mutation.   2

 Many candidates find describing the effects of a nonsense mutation tricky – it's essential to remember that these mutations affect stop codons.

➡ **Model answers and commentary can be found on page 47.**

What you need to know about chromosome mutations

Chromosome structure mutations include **duplication**, **deletion**, inversion and **translocation**.

Key diagram

Summary of the effects of mutation on chromosome structure

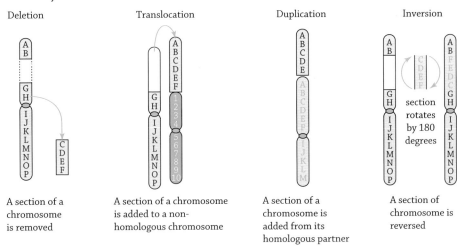

Deletion

A section of a chromosome is removed

Translocation

A section of a chromosome is added to a non-homologous chromosome

Duplication

A section of a chromosome is added from its homologous partner

Inversion

A section of chromosome is reversed

C-type question

13. Name **two** examples of structural chromosome mutations.   2

A-type question

14. Describe how **two** named chromosome mutations affect the structure of the chromosomes involved.   2

➡ **Model answers and commentary can be found on page 48.**

What you need to know about duplication mutations and evolution

Duplication allows potential beneficial mutations to occur in a duplicated gene while the original gene can still be expressed to produce its protein.

Key diagram

A duplication might produce extra copies of an important gene, which can then be further mutated while the original gene can still be expressed to produce its protein

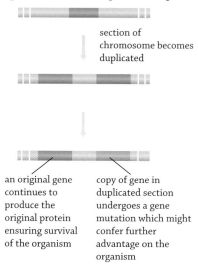

section of chromosome becomes duplicated

an original gene continues to produce the original protein ensuring survival of the organism

copy of gene in duplicated section undergoes a gene mutation which might confer further advantage on the organism

C-type question

15. Name the type of chromosome mutation which produces copies of genes which can undergo mutation to produce a new protein.   1

A-type questions

16. Explain the importance of mutations in the process of evolution.   2
17. Explain the role of duplication in the process of evolution.   2

Extended response questions

18. Name and describe the types of structural mutation of chromosomes.   4

19. Write notes on mutation under the following headings:
    (a) single gene mutations   4
    (b) chromosome mutations   3

Model answers and commentary

Question		Answer	Marks	Commentary with hints and tips
1		Mutation	1	Mutations can alter genes, gene expression or chromosomes. Mutations of genes result in no protein or an altered protein being expressed.
2		A random change in the genome that can result in no protein/an altered protein being expressed	1	Remember 'ROLF': Random Occurrence and Low Frequency.
3		The alteration of a DNA nucleotide sequence	1	Single gene mutations include deletion, insertion and substitution of nucleotides.
4		Substitution; insertion; deletion **(Any 2)**	2	Remember 'SID': Substitution, Insertion and Deletion.
5	(a)	Mutation changes a codon to a stop codon, which results in a shorter polypeptide	1	'STOP that NONSENSE' might remind you that a nonsense mutation is linked to a stop codon.
	(b)	Can result in intron being left in the mature mRNA/ exon being removed	1	Make the link between RNA splicing in Chapter 3 and this mutation of the splice site and how it changes the mature mRNA.
6		Substitution may result in missense mutation; Gives minor changes to protein/results in only one amino acid being changed **OR** Substitution may result in nonsense mutation; Loss/production of a stop codon **OR** may affect a splice site and cause introns to be left in a primary transcript	2	Remember: a **mis**sense mutation results in a **mis**take in the amino acid being coded for. 'STOP that NONSENSE' might remind you that a nonsense mutation is linked to a stop codon.

7		Insertion of a nucleotide **OR** deletion of a nucleotide; All amino acids coded for after the mutation could be affected	2	Both a description and an effect are needed.
8		Single nucleotide replaced by another with a different base; Results in the loss/ production of a stop codon	2	Both a description and an effect are needed. '**STOP that NONSENSE**' might remind you that a nonsense mutation is linked to a stop codon.
9		Nucleotide removed from DNA sequence and so each triplet/codon after the mutation is affected; All amino acids coded for after the mutation could be affected	2	An explanation is needed – the idea of **all the triplets** and **all the amino acids** being affected after the point at which the mutation took place.
10		Can result in intron being left in the mature mRNA **OR** can result in an exon being removed from mature mRNA; Leading to an altered protein	2	Make the link between RNA splicing in Chapter 3 and this mutation of the splice site and how it changes the mature mRNA transcript.
11		Single nucleotide is replaced by another with a different base; Results in only one amino acid being changed	2	Both a description and an effect are needed. A **mis**sense mutation results in a **mis**take in the amino acid being coded for.
12		Single nucleotide replaced by another with a different base; Results in the loss/ production of a stop codon	2	Both a description and an effect are needed. '**STOP that NONSENSE**' might remind you that a nonsense mutation is linked to a stop codon.
13		Duplication; deletion; inversion; translocation **(Any 2)**	2	Remember '**DICTD**': **D**eletion, **I**nversion, **C**hromosome mutation, **T**ranslocation, **D**uplication.

14		Translocation; Genes/ sections of one chromosome become attached to another chromosome **OR** Deletion; Genes/sections of chromosome deleted **OR** Inversion; Genes/sections of chromosome rotate through 180°/flipped **OR** Duplication; Genes/sections of chromosome/pieces of chromosome are duplicated/ repeated **(Any 2)**	2	Both the name **and** a description are needed for each mark.
15		Duplication	1	Remember that duplication mutations are important in evolution because they provide extra copies of genes which can then mutate further while the original gene is still retained.
16		Mutation produces new genes/alleles; May provide a selective advantage	2	Mutations are the raw material for evolution. Mutations provide variation upon which natural selection can act.
17		In duplication a second copy of a gene/section of a chromosome is present; A single gene mutation in a duplicated region of a chromosome can produce an advantageous gene/ favourable characteristic without the loss of an existing gene	2	It is thought that duplicated genes can undergo single gene mutations without affecting the functioning of the original copy of the gene.

18	1	Deletion	4
	2	Involves loss of gene(s)/section of chromosome	
	3	Translocation	
	4	Involves gene(s)/section of one chromosome joining to another	
	5	Duplication	
	6	Involves gene(s)/section of one chromosome copied within the chromosome	
	7	Inversion	
	8	Section of chromosome is reversed/set of genes rotating through 180° **(Any two names = 2 and matching effects = 2)**	
19	(a) 1	(Single gene) mutations are random changes in DNA sequences/genes/alleles/the genome;	4
	2	Single gene mutation name **AND** description from: Substitution – base/base pair/nucleotide is replaced/substituted by another Insertion – base/base pair/nucleotide is added/inserted Deletion – base/base pair/nucleotide is removed/deleted	
	3	Another single gene mutation name **AND** description	
	4	Insertion/deletion results in a frameshift mutation/expansion of a nucleic acid sequence	
	5	(Single gene) mutations are important in evolution	
	6	Splice-site mutations can alter the mature mRNA **OR** result in exon removal **OR** result in introns remaining present **(Any 4)**	
	(b) 1	Chromosome mutation can involve changes to chromosome structure	3
	2	One chromosome mutation name **AND** description from: Translocation – genes/sections of chromosome from one chromosome become attached to another chromosome Deletion – genes/sections of chromosome deleted from chromosome Inversion – genes/sections of chromosome/rotate through 180°/flipped Duplication – genes/sections of chromosome/pieces of chromosome are duplicated/repeated	
	3	Another chromosome mutation name **AND** description	
	4	Another chromosome mutation name **AND** description **(Any 3)**	

7 Evolution

What you need to know about the definition of evolution

Evolution is the result of changes in organisms over generations as a result of genomic variations.

Key diagram

Evolutionary change shown by the human skull – note that the changes include the gradual enlargement of the brain case and the gradual decrease in size of the jaw

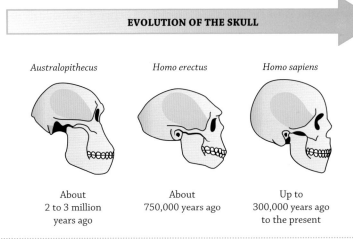

EVOLUTION OF THE SKULL

Australopithecus	*Homo erectus*	*Homo sapiens*
About 2 to 3 million years ago	About 750,000 years ago	Up to 300,000 years ago to the present

C-type question

1. Give the term used for changes to organisms over generations as result of genomic variations.   1

A-type question

2. Explain what is meant by the term 'evolution'.   1

➡ **Model answers and commentary can be found on page 56.**

What you need to know about the mechanism of evolution

Natural selection is the non-random increase in frequency of DNA sequences that increases survival and the non-random reduction in the frequency of **deleterious sequences**.

Changes in phenotype frequency can arise as a result of stabilising, directional and disruptive selection.

In **stabilising selection**, an average phenotype is selected for and extremes of the phenotype range are selected against.

In **directional selection**, one extreme of the phenotype range is selected for.

In **disruptive selection**, two or more phenotypes are selected for.

Natural selection is more rapid in prokaryotes.

Prokaryotes can exchange genetic material **horizontally**, resulting in faster evolutionary change than in organisms that only use **vertical** transfer.

Horizontal gene transfer is where genes are transferred between individuals in the same generation.

Vertical gene transfer is where genes are transferred from parent to offspring as a result of sexual or asexual reproduction.

Key diagram

Summary of the effect of different types of non-random natural selection – the shaded regions of the distributions are those phenotypes which are beneficial and survive natural selection

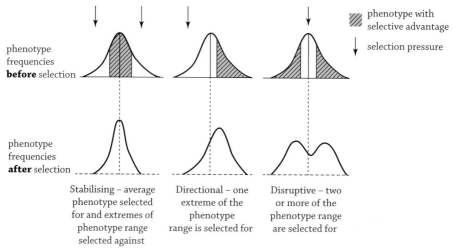

C-type questions

3. Give the term used to describe the non-random increase or decrease in genetic frequencies. 1

4. Name the form of natural selection that tends to result in the phenotype in a range becoming more aligned with a mean value. 1

5. Name the form of natural selection that tends to move the average phenotype in a range towards an extreme value. 1

6. Name the form of natural selection that tends to favour two extreme phenotypes in a range and results in two or more common phenotypes. 1

A-type questions

7. Describe natural selection in terms of the survival of the fittest. 2

8. Describe the effect of stabilising selection. 1

9. Describe the effect of directional selection. 1

10. Describe the effect of disruptive selection. 1

> **!** Many candidates struggle to explain how natural selection is involved in the evolution of bacterial populations which are resistant to antibiotics – remember that resistance starts as a mutation which is selected for and so spreads quickly through populations.

➡ **Model answers and commentary can be found on page 56.**

Key diagram

Representation of vertical and horizontal inheritance – species A, B and C have had genetic sequences passed to them vertically from parents to offspring but species B has also had a horizontal transfer of genetic sequences

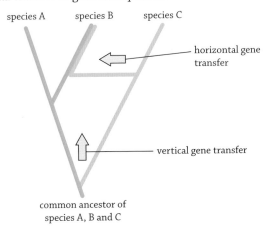

species A   species B   species C

horizontal gene transfer

vertical gene transfer

common ancestor of species A, B and C

C-type questions

11. Around one billion years ago genes were transferred between archaea and bacteria.

    Give the term that describes this form of gene transfer.   1

12. Genes can be transferred from parent to offspring as a result of sexual and asexual reproduction.

    Give the term that describes this form of gene transfer.   1

13. Prokaryotes can transfer genes to their offspring asexually.

    Give the term that describes this form of gene transfer.   1

A-type questions

14. Plasmids with antibiotic resistance genes can be passed to other bacterial species by horizontal transfer.

    Describe the process of horizontal transfer.   1

 Many candidates have problems with describing horizontal gene transfer and how it can produce rapid evolution in bacteria – remember that bacteria can take up plasmids from the environment.

15. Describe gene transfer by vertical inheritance.   2

➡ **Model answers and commentary can be found on page 56.**

What you need to know about speciation

A **species** is a group of organisms capable of interbreeding and producing fertile offspring, and which does not normally breed with other groups.

Speciation is the generation of new biological species by evolution as a result of isolation, mutation and selection.

Isolation barriers are important in preventing gene flow between sub-populations during speciation.

Geographical barriers lead to **allopatric speciation** and **behavioural** or **ecological barriers** lead to **sympatric speciation**.

Key diagram

Allopatric speciation can occur when a geographical barrier such as a new river valley isolates sub-populations; sympatric speciation occurs when an ecological or behavioural barrier caused by an initial mutation prevents gene flow between populations

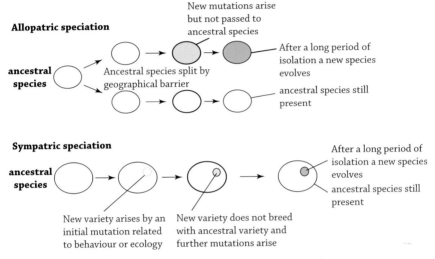

C-type questions

16. Give the meaning of the term 'speciation'.   1
17. Identify the type of speciation that involves a geographical barrier.   1
18. Identify the type of speciation that involves a behavioural or ecological barrier.   1
19. Name the type of isolation barrier involved in allopatric speciation.   1
20. Name **two** types of isolation barrier involved in sympatric speciation.   2

A-type questions

21. Describe the key features of sympatric speciation.   2
22. Describe the key features of allopatric speciation.   2
23. Explain how the results of breeding experiments between the two populations could be used to determine whether or not they were separate species.   1

Extended response questions

24. One type of selection pressure is stabilising selection.

    Give an account of this type of selection pressure and the names and effects of **two** other types of selection pressure on populations.   5

25. Give an account of allopatric and sympatric speciation.   5

26. Give an account of the roles of mutation and natural selection in the formation of new species.   7

Model answers and commentary

Question		Answer	Marks	Commentary with hints and tips
1		Evolution	1	Evolution is the result of changes in organisms over generations as a result of genomic variations. It is good to have a working definition of evolution.
2		Changes in genomic sequences in organisms that result in changes to the organisms over time/ successive generations	1	It is much harder when you need to remember and respond with the definition. As always, make flash cards for the tricky terms from each chapter.
3		Natural selection	1	Use flash cards to help you learn these terms and definitions. Do this for all the key words.
4		Stabilising selection	1	Draw thumbnail graphs showing these forms of natural selection and flash cards to help you learn these terms and definitions.
5		Directional selection	1	
6		Disruptive selection	1	
7		Organisms which are better adapted/have a favourable characteristic/selective advantage; Survive **AND** reproduce	2	You need to use these key words: favourable characteristics, selective advantage, survive **AND** reproduce.
8		Natural selection that favours a middle value of a varied characteristic	1	Giving the definition is always harder that just remembering the term.
9		Natural selection that tends to favour an extreme value of a varied characteristic	1	When the average phenotype moves/ shifts towards an extreme value in a range.
10		Natural selection that favours two different values/extreme phenotypes of a varied characteristic	1	The 'twin peak' graph.
11		Horizontal (inheritance/ gene transfer)	1	This occurs within the organisms in the same generation. No offspring are involved.
12		Vertical (inheritance/gene transfer)	1	Sexual **OR** asexual, if offspring are produced it's vertical!

13		Horizontal (inheritance/gene transfer)	1	Don't be tempted to identify vertical or horizontal inheritance patterns from the way a diagram is laid out – a horizontal arrow in a diagram does not necessarily mean horizontal inheritance! Strawberry runners would appear horizontal but offspring are produced – so it is vertical inheritance!
14		A plasmid from one bacterium is taken up from the environment by another bacterium **OR** a plasmid is passed across from one bacterial cell to another	1	Link with Chapter 15.
15		Genetic material is passed from parent to offspring; by sexual or asexual reproduction	2	Parent to offspring so sexual or asexual.
16		The evolution of two or more species from a common ancestor **OR** evolutionary process by which new species are formed	1	Use flash cards to help you learn these terms and definitions. Do this for all the key words.
17		Allopatric speciation	1	Geographical barriers include mountains, oceans and other physical features.
18		Sympatric speciation	1	Behavioural barriers are often related to courtship and other features of reproduction. Ecological barriers are often related to habitat preferences.
19		Geographical	1	Examples include mountains, rivers and continental drift.
20		Behavioural; ecological	2	Better to use these names rather than try to give examples.
21		Speciation in which gene flow is prevented; by ecological or reproductive barriers	2	Remember '**I'M A NEW SPECIES**' to represent the order of events in speciation – isolation, mutation and natural selection.
22		Speciation in which gene flow is prevented; by geographical barriers	2	Isolating barriers prevent gene exchange/gene flow/interbreeding between sub-populations.

23		New species form if populations can no longer interbreed to produce fertile young	1	This is the biological definition of a species.	
24	1	Stabilising selection favours individuals with a central value in the range of variation			5
	2	Directional selection			
	3	Favours individuals with characteristics at one extreme of the range			
	4	Disruptive selection			
	5	Favours individuals at different values in a range **OR** acts against individuals in the middle of a range **(All 5)**			
25	1	Isolation barriers act as barriers to gene flow/gene exchange/ interbreeding between sub-populations of a species			5
	2	In allopatric speciation, gene flow is prevented by a geographical barrier			
	3	In sympatric speciation, gene flow is prevented by a behavioural barrier			
	4	In sympatric speciation, gene flow is prevented by an ecological barrier			
	5	Different mutations in each sub-population			
	6	Natural selection is different for each sub-population			
	7	Speciation results when sub-populations can no longer interbreed to produce fertile offspring **(Any 5)**			
26	1	Mutation produces variation within species			7
	2	Mutation can be deleterious/harmful or beneficial			
	3	There is a struggle for survival/competition between individuals of the species			
	4	Those which are fittest/best suited/with beneficial mutations/selective advantage survive			
	5	The survivors reproduce			
	6	Survivors pass mutations/favourable genes/alleles/characteristics to their offspring			
	7	After long periods of time new species may form			
	8	New species form/speciation occurs when sub-populations can no longer interbreed to produce fertile offspring **(Any 7)**			

8 Genomic sequencing

What you need to know about sequence data

In genomic sequencing the sequence of nucleotide bases can be determined for individual genes and entire genomes.

Computer programs can be used to identify base sequences by looking for sequences similar to known genes.

To compare **sequence data**, computer and statistical analyses (**bioinformatics**) are required.

Comparison of genomes reveals that many genes are highly conserved across different organisms.

Many genomes have been sequenced, particularly of disease-causing organisms, pest species and species that are important model organisms for research.

Key diagram

Bands representing the gene sequences in the genomes of different species; these can be analysed using computer and statistical analyses (bioinformatics)

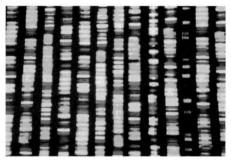

C-type questions

1. Name the type of data which can be used to compare the genomes of different species.   1
2. Give the term applied to the use of computers and statistics in the analysis of sequence data.   1

A-type questions

3. Describe the information obtained by the process of genomic sequencing.   1
4. Describe how bioinformatics can be used to work out the evolutionary relatedness of two species.   2

➡ **Model answers and commentary can be found on page 63.**

Key diagram

The genomes of human, chimpanzee and gorilla, showing many conserved gene sequences

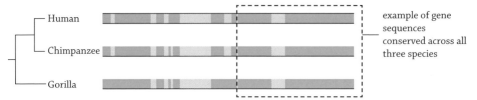

C-type question

5.   Give **two** examples of the types of species for which genomes have been sequenced.   2

A-type question

6.   Explain what is meant by the statement that much of the genome is highly
     conserved across different species.   1

➡ **Model answers and commentary can be found on page 64.**

What you need to know about phylogenetics and molecular clocks

Phylogenetics is the study of evolutionary history and relationships.

Evidence from phylogenetics and **molecular clocks** has been used to determine the main sequence of events in evolution.

The sequence of events can be determined using sequence data and **fossil evidence**.

Sequence data is used to study the evolutionary relatedness among groups of organisms.

Sequence divergence is used to estimate the time since lineages diverged.

Comparison of sequences provides evidence of the three domains of life – **bacteria**, **archaea** and eukaryotes.

Sequence data and fossil evidence have been used to determine the main sequence of events in the evolution of life: cells, last universal ancestor, prokaryotes, photosynthetic organisms, eukaryotes, multicellularity, animals, vertebrates, land plants.

Molecular clocks are used to show when species diverged during evolution.

Molecular clocks assume a constant mutation rate and show differences in DNA sequences or amino acid sequences.

The differences in sequence data between species indicate the time of divergence from a common ancestor.

Key diagram

Generalised phylogenetic tree diagram showing divergences from common ancestors and evolutionary relatedness between species

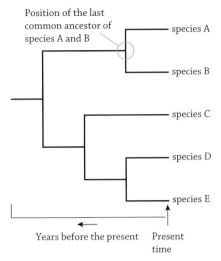

C-type questions

7. Give the name for the study of evolutionary relatedness.   1

8. State **two** sources of evidence which allow phylogenetic trees to be constructed.   2

9. Name the three main domains of life.   1

A-type question

10. From the phylogenetic tree shown in the **Key diagram** above, give the evidence which confirms that species A and B are more closely related to each other than to species C.   1

 Many candidates have difficulty in interpreting phylogenetic trees – make sure you know where the common ancestors are located.

➡ **Model answers and commentary can be found on page 64.**

Key diagram

The approximate times at which various events in the evolution of life occurred, as determined using sequence data and fossil evidence

Event in evolution	Relative time
Appearance of cells	Earliest event
Existence of the last universal ancestor (LUA)	
Appearance of prokaryotic cells	
Appearance of photosynthetic organisms	
Appearance of eukaryotic cells	
Appearance of multicellular organisms	
Appearance of animals	Most recent event
Appearance of vertebrate animals	
Appearance of land plants	

C-type question

11. Name the types of data which have been used to determine the main sequence of events in the evolution of life. 2

➡ **Model answers and commentary can be found on page 64.**

Key diagram

A molecular clock for haemoglobin which shows how differences in its amino acid sequence can be used to determine the times at which various vertebrate groups diverged

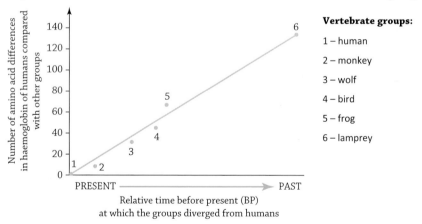

Vertebrate groups:

1 – human

2 – monkey

3 – wolf

4 – bird

5 – frog

6 – lamprey

C-type question

12. Name the type of graph that shows differences in sequence data for a protein against time. 1

A-type questions

13. Describe the assumption which is made in the construction of a molecular clock used to show the divergence of species. 1

14. Describe the information which is needed to construct a molecular clock. 1

A tricky idea to understand is that molecular clocks are based on the assumption that mutation rates are constant.

➡ **Model answers and commentary can be found on page 64.**

What you need to know about personalised medicine

An individual's genome can be analysed to predict the likelihood of developing certain diseases.

Pharmacogenetics is the use of genome information in the choice of drugs.

An individual's personal genome sequence can be used to select the most effective drugs and dosage to treat their disease (**personalised medicine**).

Key diagram

An individual and their genome

C-type questions

15. Give the meaning of the term 'pharmocogenetics'.	1
16. Give **one** possible benefit of personalised medicine to patients in the future.	1

Extended response questions

17. Give an account of personalised genomics and medicine.	4
18. Give an account of genomic sequencing under the following headings:	
(a) phylogenetics and molecular clocks	6
(b) personal genomics and health	3

Model answers and commentary

Question		Answer	Marks	Commentary with hints and tips
1		Gene sequences/genomic sequencing **OR** differences in sequence data for a protein/ sequence of amino acids in a protein	1	The construction of biological databases allows the comparison of the order of the nucleotides/bases or the order of amino acids in proteins.
2		Bioinformatics	1	The use of computers to interpret sequence data.
3		The sequence of DNA nucleotide bases can be determined for individual genes/entire genomes **OR** differences in sequence data for a protein/ sequence of amino acids in a protein	1	Remember sequence = order. In Chapter 3, we stressed that the sequence of the bases determines the sequence of amino acids in a protein.

4		Computer and statistical analysis; To compare sequence data	2	Computers allow the vast quantities of information held in the databases to be compared.
5		Disease-causing organisms; pest species; model organisms for research **(Any 2)**	2	These are the three examples you are expected to know.
6		Much of the genome is the same in many different species	1	Some genes, such as those which express respiration enzymes, are found in nearly all living organisms – they have been maintained by natural selection and so are highly conserved.
7		Phylogenetics	1	The relationships of one organism to another according to evolutionary similarities and differences.
8		Sequence data; fossil evidence	2	Data includes sequences of nucleotides/bases and amino acid sequences in specific proteins. Fossils help to confirm accurate dating.
9		Bacteria **AND** archaea **AND** eukaryotes	1	The bacteria and archaea are prokaryotic microorganisms, or single-celled organisms whose cells have no nucleus.
10		The common ancestor of species A and B is more recent than their common ancestor with species C	1	The more closely related two species are, the more recent their common ancestor existed.
11		Sequence data; fossil evidence	2	Combining evidence from both sources is needed to work out the main sequence of events in evolution.
12		Molecular clock	1	Uses the mutation rate of biomolecules to deduce the time in prehistory when two or more life forms diverged. The biomolecular data used for such calculations are usually nucleotide sequences for DNA or amino acid sequences for proteins.
13		That rates of mutation are constant	1	Assuming a constant rate allows the construction of molecular clocks.

14		Difference in nucleotide/base sequences between genes **OR** amino acid sequences between proteins from different species	1	It is these differences which allow the different species to be placed on the molecular clock.
15		The use of genome information in the choice of drugs	1	Individuals with different genomes may respond to different drugs in different ways.
16		Provides more information on the likelihood of a treatment being successful in a specific individual **OR** Bespoke drugs/medication/treatment based on individual's genome **OR** Type/dose of drug based on individual's genome **OR** Genomic differences influence the effectiveness of treatments/drugs **OR** Different treatments can be designed for an individual	1	This is the study of inherited genetic differences in drug metabolic pathways. This can affect how an individual responds to drugs, both in terms of therapeutic effect as well as an adverse effect. How the genetic makeup of an individual affects their response to drugs.
17	1	Personalised genomics is having an individual's genome sequenced	4	
	2	Analysis of an individual genome could lead to personalised medicine		
	3	Genetic components of disease could be revealed		
	4	The likelihood of success of treatments could be estimated		
	5	Selection of most effective drug		
	6	Selection of most effective dosage **(Any 4)**		
18	(a) 1	Genomics involves the study of gene sequences	6	
	2	Gene sequences are used to show evolutionary relatedness		
	3	Evolutionary relatedness is the basis of phylogenetic trees		
	4	To add timescales to phylogenetic trees, fossils are needed		
	5	Molecular clocks are based on sequence differences of a particular protein		
	6	Differences in sequences related to a protein in different species are graphed on one axis		
	7	The other axis shows the timescale of divergence based on relative sequence differences **(Any 6)**		
	(b) 1	Personal genomics involves analysing the genome of an individual	3	
	2	Disease risk has a genetic component		
	3	Genomic differences influence the effectiveness of treatments/drugs		
	4	So different treatments can be designed for an individual **(Any 3)**		

Metabolic pathways

What you need to know about metabolic pathways

Metabolic pathways are integrated and controlled pathways of enzyme-catalysed reactions within a cell.

Metabolic pathways can have reversible steps, irreversible steps and alternative routes.

Reactions within metabolic pathways can be **anabolic** or **catabolic**.

Anabolic reactions build up large molecules from small molecules and require energy.

Catabolic reactions break down large molecules into smaller molecules and release energy.

Key diagram

Simple metabolic pathway with one reversible and several irreversible steps and an alternative route to the final metabolic product E

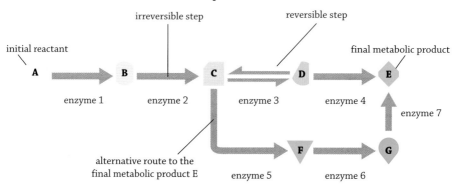

C-type questions

1. State what is meant by a metabolic pathway. 1
2. Give the term used to describe chemical reactions that require an input of energy. 1
3. Give the term used to describe chemical reactions that involve the breakdown of complex molecules, resulting in the release of energy. 1
4. State **two** differences between anabolic and catabolic pathways. 2

A-type questions

5. Describe the difference between a reversible and an irreversible step. 2
6. Explain what is meant by an alternative route in a metabolic pathway. 1

⟹ **Model answers and commentary can be found on page 70.**

What you need to know about proteins in membranes

Proteins embedded in phospholipid membranes have functions such as pores, pumps or enzymes.

Key diagram

Eukaryotic cells contain many membrane-bound organelles, including mitochondria and chloroplasts. These membranes contain proteins which are pores, pumps and enzymes.

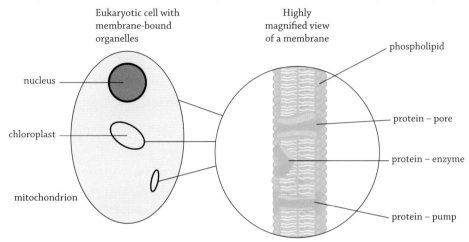

C-type questions

7. Name the **two** chemical components of membranes.   2
8. Give **two** roles of proteins embedded in phospholipid membranes.   2
9. Give **two** examples of organelles bounded by membranes.   2

A-type questions

10. Name an enzyme which is embedded in the inner mitochondrial membrane.   1

⟹ **See 'What you need to know about the electron transport chain' in Chapter 10.**

11. Describe the roles of protein pumps embedded in the inner mitochondrial membrane.   1

⟹ **See 'What you need to know about the electron transport chain' in Chapter 10.**

⟹ **Model answers and commentary can be found on page 71.**

What you need to know about control of metabolic pathways

Metabolic pathways are controlled by the presence or absence of particular enzymes and the regulation of the rate of reaction of key enzymes.

Induced fit occurs when the **active site** of the enzyme changes shape to better fit the substrate after the substrate binds.

The concentration of substrate and end product affect the direction and rate of an enzyme reaction.

As the substrate concentration increases, the rate of the enzyme reaction increases until all of the active sites are occupied by the substrate.

Substrate molecule(s) have a high affinity for the active site and the subsequent products have a low affinity, allowing them to leave the active site.

Some metabolic reactions are reversible and the presence of a substrate or the removal of a product will drive a sequence of reactions in a particular direction.

Competitive inhibitors bind at the active site preventing the substrate from binding.

Competitive inhibition can be reversed by increasing substrate concentration.

Non-competitive inhibitors bind away from the active site but change the shape of the active site, preventing the substrate from binding.

Non-competitive inhibition cannot be reversed by increasing substrate concentration.

Feedback inhibition occurs when the end product in the metabolic pathway reaches a critical concentration then inhibits an earlier enzyme, blocking the pathway, and so prevents further synthesis of the end product.

Key diagram

Enzyme affinity and induced fit mechanism of a catabolic reaction

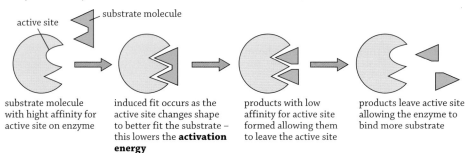

| substrate molecule with hight affinity for active site on enzyme | induced fit occurs as the active site changes shape to better fit the substrate – this lowers the **activation energy** | products with low affinity for active site formed allowing them to leave the active site | products leave active site allowing the enzyme to bind more substrate |

C-type questions

12. Name the molecules in cells that reduce the activation energy needed for a chemical reaction to occur.   1

13. State the effect an enzyme has on the activation energy of an enzyme-catalysed reaction.   1

A-type questions

14. Describe the role of genes in the control of metabolic pathways.   2

➡ **See Chapter 3.**

15. Describe the role of the active site of an enzyme during a reaction.   2

16. Describe the mechanism of induced fit.   2
17. Describe the effect of an increase in substrate concentration on the direction
    and rate of an enzyme reaction.   2
18. Explain how enzymes speed up the rate of reactions in metabolic pathways.   2
19. Describe the effect of substrate concentration and end-product concentration
    on a reversible metabolic reaction.   2

➡ **Model answers and commentary can be found on page 71.**

Key diagram

An enzyme molecule, showing the active site and the non-competitive inhibitor binding site

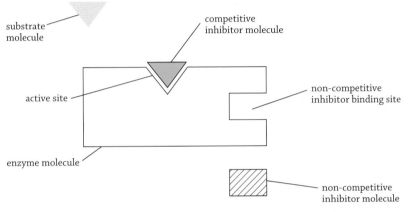

Technique

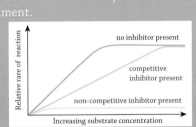

Altering reaction rates by changing substrate concentration or using inhibitors is a technique you need to be familiar with for your exam and one that could be used in your assignment.

This graph shows the effect on reaction rate of changing substrate concentration in the presence of different inhibitors and is the basis of many exam questions.

C-type questions

20. Name the type of inhibition in which the inhibitor molecule has a similar shape to
    the substrate molecule.   1
21. Name the type of inhibition in which the inhibitor molecule binds to a site other
    than the active site and changes the shape of the active site.   1
22. Name the type of inhibition in which an enzyme early in the metabolic pathway is
    inhibited by the final substance produced.   1

A-type questions

23. Describe the effect of increasing substrate concentration on the rate of reaction in the presence of a competitive inhibitor.   2

24. Describe the effect of increasing substrate concentration on the rate of reaction in the presence of a non-competitive inhibitor.   2

25. Explain the advantage of end-product inhibition to a cell.   1

26. Explain the difference between a competitive inhibitor and non-competitive inhibitor.   2

27. Explain the effect of a competitive inhibitor on the rate of an enzyme reaction.   2

28. Explain the effect of a non-competitive inhibitor on the rate of an enzyme reaction.   2

Extended response questions

29. Describe and compare anabolic and catabolic reactions.   4

30. Write notes on the functions of proteins embedded in membranes.   3

31. Give an account of the competitive and non-competitive inhibition of enzymes.   4

32. Give an account of the induced fit model of enzyme action.   4

33. Give an account of the structure and mode of action of enzymes.   6

Model answers and commentary

Question	Model answer	Marks	Commentary with hints and tips
1	A series/sequence of chemical reactions controlled by enzymes	1	Each stage in the metabolic pathway is controlled by a specific enzyme coded for by a gene.
2	Anabolic	1	The building up of complex molecules from simpler substances. Examples include protein synthesis and DNA replication.
3	Catabolic	1	The breakdown of complex molecules into simpler substances. Examples include glycolysis and the citric acid cycle.
4	Anabolic pathways involve the building up of complex molecules from simpler substances and catabolic pathways involve the breakdown of complex molecules into simpler substances; Anabolic pathways require the input of energy and catabolic pathways usually release energy	2	Make flash cards of these terms. Make up a pack for each chapter and have someone test you on them. Practice makes perfect and will definitely gain you additional marks.

5		With a reversible step the product(s) can react to form the original substrate/reactant(s); In an irreversible reaction the reactants react to form the products, which cannot revert back into reactants	2	Check out the process of glycolysis in Chapter 10, which has examples of both.
6		When a starting substrate/ intermediate is acted upon by a different enzyme and undergoes/ follows a different pathway/ bypasses the other steps	1	Alternative = different.
7		Phospholipid molecules; proteins	2	Components = parts.
8		Pores; pumps; enzymes **(Any 2)**	2	Remember 'Poly**PEP**tides': proteins for **p**ores, **e**nzymes and **p**umps.
9		Mitochondria; chloroplasts; nucleus **(Any 2)**	2	The organelles in a cell are membrane-bound compartments in which metabolic pathways are localised.
10		ATP synthase	1	Note that the name of this enzyme can be found in Chapter 10, 'What you need to know about the electron transport chain'.
11		Use energy to pump hydrogen ions (across the inner mitochondrial membrane)	1	Note that the description of this process can be found in Chapter 10, 'What you need to know about the electron transport chain'.
12		Enzymes	1	The energy required to initiate a chemical reaction is called the activation energy.
13		Lowers/decreases the activation energy	1	High temperatures often supply the activation energy in non-living situations, but in cells enzymes reduce the activation energy needed for a reaction to occur.

14		Each step in a metabolic pathway is controlled by a specific enzyme; Each enzyme is coded for by a gene; The order of bases in the gene determines the order of amino acids, which determines structure/shape/function of the protein/enzyme **OR** Genes code for the enzymes; that control the metabolic pathway/chemical reactions	2	Metabolic pathways are controlled by enzymes. Enzymes are proteins. Genes code for proteins and so code for the enzymes that control the metabolic pathway. See Chapter 3.
15		The active site has a high affinity/ attraction for the substrate molecule and a low affinity for the product; This lowers the activation energy of the chemical reaction	2	Affinity is a good term to use when talking about the attraction a substrate has for an active site.
16		The enzyme is flexible and so the active site can change shape; The substrate induces/causes the active site to change shape; The active site can orientate/ alter the position of the substrate molecules so that they fit more closely	2	This orientates the reactants into the correct positions for the reaction to take place. This results in a lowering of the activation energy.
17		Increase in substrate concentration drives the chemical reaction in the direction of the end product; Increases the rate of reaction	2	This is only tricky because it may not have been emphasised in class and so is less well remembered.
18		The active site can orientate/ alter the position of the substrate molecules so that they fit more closely; The activation energy is lowered when an enzyme is involved	2	A rewording of Question 17, with a similar answer. You need to be able to recognise the different ways of asking the same question.
19		Increase in substrate concentration drives the chemical reaction in the direction of the end product; Increase in the end product concentration drives the chemical reaction in the direction of the substrate	2	A reversible chemical reaction is one that can go in both directions. The reactants can turn into products, and products can turn back into reactants. Remember that reversible reactions are shown by double arrows ($\rightleftharpoons$).

20		Competitive inhibition	1	Competitive inhibition can be reduced by increasing the substrate concentration.
21		Non-competitive inhibition	1	Since the shape of the active site has been changed, increasing the substrate concentration will not increase the reaction rate.
22		Feedback inhibition	1	This process stops the metabolic pathway until the end product concentration decreases.
23		Reaction rate increases with substrate concentration; But as substrate concentration increases further there is no further effect on rate of reaction	2	A description is all that is needed, but both parts need to be included for 2 marks.
24		Inhibitor reduces rate of reaction nearly to zero; increasing substrate makes no difference	2	A description is all that is needed, but both parts need to be included for 2 marks.
25		Prevents/stops the production of too much end product **OR** Controls/regulates concentration of end product **OR** Ensures efficient use of resources/ substrate	1	Occurs when an end product inhibits the activity of an enzyme that catalysed a reaction earlier in the pathway that produced it.
26		Competitive binds reversibly at the active site; Non-competitive binds irreversibly at another binding site	2	Explanation based on binding sites and reversibility differences.
27		The rate of the enzyme reaction decreases/is reduced; Competitive inhibitor has a similar shape to the substrate molecule and attaches to the active site of the enzyme	2	An 'explain' question which needs both the effect on the reaction rate **and** the reason.
28		The rate of reaction is reduced to nearly zero; Inhibitor has caused a change in the shape of the active site so that substrate cannot bind	2	An 'explain' question which needs both the effect on the reaction rate **and** the reason.

29	1	An anabolic reaction is a synthesis/build-up reaction **OR** Where simple molecules are built up into more complex/large molecules	4
	2	Anabolic reactions require the input/take up of energy/ATP	
	3	A catabolic reaction is the breakdown/degradation of molecules/ substances **OR** A catabolic reaction is where complex/large molecules are changed into more simple molecules	
	4	Energy/ATP is released/given off in a catabolic reaction	
	5	Both can have reversible/irreversible steps	
	6	Both can have alternative routes **(Any 4)**	
30	1	Pores allow diffusion of small molecules	3
	2	Pumps/carrier proteins actively move substances/ions against their concentration gradient	
	3	Some proteins act as enzymes **(All 3)**	
31	1	Competitive inhibitors block active sites	4
	2	Competitive inhibitors slow down rates of reaction	
	3	The effects of competitive inhibitors are reduced by increasing substrate concentration	
	4	Non-competitive inhibitors bind to site other than active site	
	5	Non-competitive inhibitors slow down reactions irreversibly	
	6	Non-competitive inhibitors are not affected by substrate concentration **(Any 4)**	
32	1	Substrate and active site may not bind initially	4
	2	Substrate has a high affinity for the active site	
	3	Active site is flexible **OR** A change in the active site occurs	
	4	Shape change induced by the binding of a specific substrate to the active site	
	5	Activation energy reduced **(Any 4)**	
33	1	Substrate has an affinity for the active site	6
	2	Induced fit model of enzyme action	
	3	Active site orientates the reactants so that they fit more closely	
	4	Enzymes lower the activation energy	
	5	Products have a low affinity for the active site	
	6	Substrate/product concentration affects the **direction** of the reactions **OR** Increasing the substrate concentration increases/ speeds up/drives forward the **rate** of the reaction	
	7	Substrate/product concentration affects the rate of the reactions **(Any 6)**	

Cellular respiration

What you need to know about the stages of cellular respiration

Glycolysis is the breakdown of glucose to pyruvate in the cytoplasm.

ATP is required for the **phosphorylation** of glucose and **intermediates** during the energy investment phase of glycolysis.

The generation of more ATP during the energy pay-off stage results in a net gain of ATP.

In aerobic conditions, **pyruvate** is broken down to an **acetyl group** that combines with **coenzyme A** forming acetyl coenzyme A.

In the **citric acid cycle** the acetyl group from acetyl coenzyme A combines with **oxaloacetate** to form **citrate**.

Citrate is gradually converted back into oxaloacetate during a series of enzyme-controlled steps, which results in the generation of ATP and release of carbon dioxide.

The citric acid cycle occurs in the matrix of the mitochondria.

In the citric acid cycle and glycolysis, **dehydrogenase** enzymes remove hydrogen ions and **electrons** and pass them to the coenzyme **NAD**, forming NADH.

The hydrogen ions and electrons from NADH are passed to the **electron transport chain** on the inner mitochondrial membrane.

Key diagram

Stages of glycolysis, which is the breakdown of glucose to pyruvate in the cytoplasm

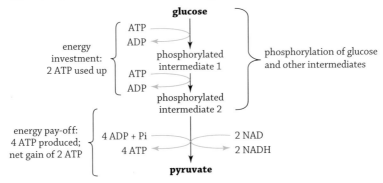

C-type questions

1. State the location of glycolysis.   1

2. Name the final product of glycolysis.   1

3. Name the substance required for the phosphorylation reactions during glycolysis.   1

4. Name the coenzyme in glycolysis which carries hydrogen to the electron transport chain.   1

5. Name the stage in respiration that has both an energy investment stage and an energy pay-off stage.   1

6. Name the enzyme which removes hydrogen ions and high-energy electrons from respiratory substrates.   1

A-type questions

7. Explain why the phosphorylation of intermediates in glycolysis is referred to as an energy investment stage.   2

8. Describe the role of the dehydrogenase enzymes in glycolysis and the citric acid cycle.   1

9. Explain the role of ATP during glycolysis.   2

10. Explain why the ATP produced in glycolysis is referred to as an energy pay-off.   2

Many candidates find describing glycolysis difficult – remember substrates, products and the different energy phases.

➡ **Model answers and commentary can be found on page 80.**

Key diagram

Progress of pyruvate under aerobic conditions in the matrix of a mitochondrion

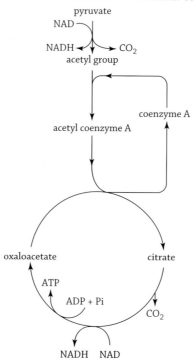

C-type questions

11. Name the stage in respiration in which citrate is synthesised from oxaloacetate and an acetyl group. 1
12. Name the molecule produced when oxaloacetate combines with acetyl coenzyme A. 1
13. Name the exact location of the citric acid cycle. 1
14. Name the hydrogen carrier that links the citric acid cycle to the electron transport chain. 1
15. Name the exact location of the electron transport chain in cells. 1

A-type question

16. Describe the role of the coenzyme NAD. 1

 Many candidates find describing the role of enzymes and NAD in the citric acid cycle difficult – remember this is all about hydrogen ions and electrons being removed and transported.

➡ **Model answers and commentary can be found on page 81.**

What you need to know the electron transport chain

The electron transport chain is a series of carrier proteins attached to the inner mitochondrial membrane.

Electrons are passed along the electron transport chain releasing energy, which allows hydrogen ions to be pumped across the inner mitochondrial membrane.

The flow of these hydrogen ions back through the membrane protein ATP synthase results in the production of ATP.

Finally, hydrogen ions and electrons combine with oxygen to form water.

Key diagram

The role of the electron transport chain in ATP synthesis

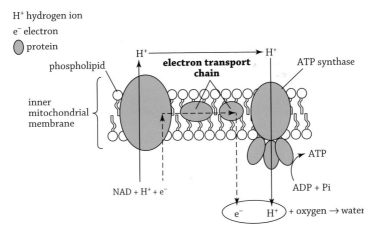

C-type questions

17. Name the location of the electron transport chain.   1
18. Name the enzyme embedded in the inner membrane of a mitochondrion responsible for the regeneration of ATP.   1
19. State the source of energy required to pump hydrogen ions across the inner mitochondrial membrane.   1
20. Identify the high-energy molecule that is produced when energy from the electron transport chain pumps hydrogen ions across the inner mitochondrial membrane. 1
21. State the role of oxygen in the electron transport chain.   1

A-type questions

22. Describe how hydrogen ions cause the synthesis of ATP during respiration.   1
23. Describe the role of the electrons that pass down the electron transport chain.   1
24. Describe the role of proteins in the electron transport chain.   2
25. Describe the role played by oxygen in the electron transport chain.   2

➡ **Model answers and commentary can be found on page 81.**

What you need to know about fermentation

In the absence of oxygen, **fermentation** takes place in the cytoplasm.

In animal cells, pyruvate is converted to **lactate** in a reversible reaction.

In plants and yeast, ethanol and CO_2 are produced in an irreversible reaction.

Fermentation results in much less ATP being produced than in aerobic respiration.

Key diagram

Fermentation in the cytoplasm of animal, plant and yeast cells

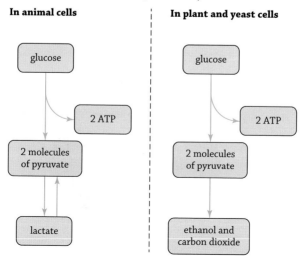

C-type questions

26. Name the respiratory pathway which occurs in the absence of oxygen.   1

27. State the location of fermentation.   1

28. Name the product of fermentation in animal cells.   1

29. Name the products of fermentation in plant and yeast cells.   2

A-type question

30. Compare the products of fermentation in animal and yeast cells.   2

➡ **Model answers and commentary can be found on page 82.**

What you need to know about the role of ATP

ATP is used to transfer energy in cellular processes which require energy.

Key diagram

Summary of the energy transfer system in cells

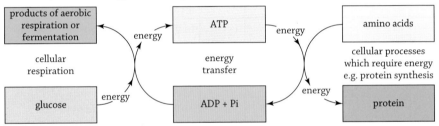

Technique

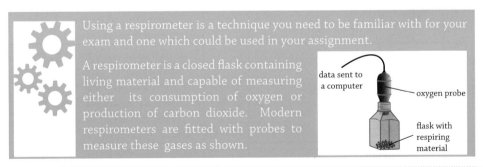

Using a respirometer is a technique you need to be familiar with for your exam and one which could be used in your assignment.

A respirometer is a closed flask containing living material and capable of measuring either its consumption of oxygen or production of carbon dioxide. Modern respirometers are fitted with probes to measure these gases as shown.

data sent to a computer

oxygen probe

flask with respiring material

C-type question

31. Name the substance used to transfer energy in cells.   1

A-type questions

32. Describe the role of ATP in cell metabolism.   1

33. Name **two** cellular processes which are anabolic.   2

➡ **See Chapters 2, 3 and 9.**

Extended response questions

34. Describe the energy investment and energy pay-off phases in glycolysis. 4
35. Give an account of the citric acid cycle in respiration. 6
36. Give a description of the location and function of the enzyme ATP synthase. 4
37. Compare fermentation in animal cells and yeast cells. 4
38. Give an account of glycolysis and the citric acid cycle in respiration. 9
39. Give an account of the electron transport chain. 8

Model answers and commentary

Question		Model answer	Marks	Commentary with hints and tips
1		Cytoplasm	1	Glyco = sugar/glucose Lysis = to split The breakdown of glucose to pyruvate.
2		Pyruvate	1	Glucose → intermediate 1 → intermediate 2 → pyruvate
3		ATP	1	In glycolysis, the phosphorylation of glucose and intermediate 1 by ATP is an energy investment phase. ATP is broken down into ADP and Pi. The Pi is then added to the glucose and then intermediate 1.
4		NAD	1	NAD + H$^+$ + electron → NADH NADH transports both the hydrogen ion and electron to the electron transport chain.
5		Glycolysis	1	Two ATP molecules are used to phosphorylate intermediates in the energy investment phase; four ATP molecules are produced in the pay-off stage. This gives a net (overall) gain of two ATP.
6		Dehydrogenase	1	Rewording of Question 1, made a little trickier by mentioning a 'group of enzymes' and using the term 'respiratory substrate' rather than glucose.
7		Two ATP molecules are used to phosphorylate intermediates in the energy investment phase; Four ATP molecules are produced in the pay-off stage	2	This gives a net (overall) gain of two ATP.

8		Dehydrogenase enzymes remove hydrogen ions **and** electrons.	1	A rewording of Question 3 but referring to named stages.
9		In glycolysis, ATP is used in the phosphorylation; of glucose **and** intermediate 1/ intermediates **OR** ATP is broken down into ADP and Pi; The Pi is then added to the glucose and then intermediate 1/intermediates	2	In glycolysis, the phosphorylation of glucose and intermediate 1 by ATP is an energy investment phase.
10		Four ATP molecules are produced in the pay-off stage; This gives a net (overall) gain of two ATP	2	Two ATP molecules are used to phosphorylate intermediates in the energy investment phase; four ATP molecules are produced in the pay-off stage. This gives a net (overall) gain of two ATP.
11		The citric acid cycle	1	If oxygen is available, pyruvate progresses to the citric acid cycle. This stage takes place in the central matrix of the mitochondria.
12		Citrate	1	The start of the citric acid cycle.
13		The central matrix of the mitochondria	1	Remember to mention the matrix in your answer. It is not enough to just say 'the mitochondrion'.
14		NAD	1	This coenzyme transports hydrogen ions **and** electrons to the electron transport chain.
15		The inner membrane of the mitochondria	1	The question asked for the exact location and did not mention the mitochondria, so again it has to be specified in your answer.
16		It transports hydrogen ions and electrons; to the electron transport chain (on the inner mitochondrial membrane)	1	$NAD + H^+ + electron \rightarrow NADH$.
17		Inner membrane of the mitochondria	1	Mitochondria with an increased number of folds produce more ATP.
18		ATP synthase	1	This question may be asked as 'Name the enzyme' **OR** 'Name the protein'.
19		Electrons	1	Remember that flowing electrons carry energy.

20		ATP	1	ATP is regenerated as the hydrogen ions flow back through the ATP synthase. It is good to use the term 'regenerated' when writing about ATP synthesis.
21		Acts as the final hydrogen ion and electron acceptor	1	Combines with hydrogen ions **and** electrons to form water.
22		ATP is regenerated as the hydrogen ions flow back through the ATP synthase	1	It is good to use the term 'regenerated' when writing about ATP synthesis.
23		Release/provide energy to pump the hydrogen ions across the inner membrane of the mitochondria	1	The moving electrons give the name to the stage of respiration – the electron transport chain.
24		The electrons cascade down the chain of proteins releasing energy; To pump hydrogen ions across the inner membrane of the mitochondria	2	Energy pumps the hydrogen ions through the protein pump in the inner membrane.
25		Acts as the final electron acceptor; Combines with hydrogen ions **and** electrons to form water	2	When describing the role of oxygen, it's better to mention water as the final product of aerobic respiration.
26		Fermentation	1	The only acceptable answer.
27		Cytoplasm	1	Remember that the cytoplasm has a role in aerobic respiration as well – glycolysis occurs there.
28		Lactate	1	Remember that the production of lactate is a reversible reaction.
29		Ethanol; CO_2	2	Remember that the production of ethanol and CO_2 in plants and yeast is an irreversible reaction.
30		Fermentation in animal cells produces lactate but in yeast produces ethanol and CO_2; Fermentation in animal cells is reversible but in yeast it is irreversible	2	Remember to have a full comparison for each point you make, mentioning both animal cells **and** yeast cells.
31		ATP	1	ATP is used to transfer the energy from cellular respiration to anabolic pathways where energy is required.

32		ATP is used to transfer the energy from cellular respiration to cellular processes which require energy	1	The word 'transfer' is key here.	
33		Protein synthesis; transcription/translation/ DNA replication; others	2	Knowledge of ATP from other chapters is required here.	
34	1	The first phase of glycolysis is called an energy investment phase			4
	2	Two ATP molecules are used up to phosphorylate glucose and intermediate 1/intermediates			
	3	Intermediate 2 is converted/broken down into two pyruvate molecules			
	4	Four ATP are produced			
	5	This results in a net/overall gain of two ATP **(Any 4)**			
35	1	If oxygen is available/in aerobic conditions pyruvate progresses to the citric acid cycle			6
	2	Pyruvate is converted/broken down to an acetyl group			
	3	The acetyl group combines with coenzyme A			
	4	Acetyl (coenzyme A) combines with oxaloacetate to form citrate			
	5	Citric acid cycle is enzyme controlled/involves dehydrogenases			
	6	ATP generated/synthesised/produced/released at substrate level in the citric acid cycle			
	7	Carbon dioxide is released from the citric acid cycle			
	8	NAD/NADH transports electrons **and** hydrogen ions to the electron transport chain **(Any 6)**			
36	1	Enzyme/protein located in the inner membrane of a mitochondrion			4
	2	Also found in the membranes of chloroplasts			
	3	Catalyses the synthesis of ATP from ADP and Pi			
	4	Hydrogen ions flow through the enzyme/ATP synthase			
	5	Flow of hydrogen ions provides the energy to drive the enzyme			
	6	Movement/flow of hydrogen ions is down the concentration gradient/ from a region of high concentration to a region of low concentration **(Any 4)**			
37	1	Fermentation in animal cells produces lactate but in yeast produces ethanol and CO_2			4
	2	Fermentation in animal cells is reversible but in yeast it is irreversible			
	3	Both occur in the absence of oxygen			
	4	Both produce less ATP than aerobic respiration			
	5	Both occur in the cytoplasm **(Any 4)**			

38	1	Glycolysis is the breakdown of glucose to pyruvate	9
	2	Two ATP molecules are used to phosphorylate intermediates in glycolysis	
	3	This is called an energy investment phase	
	4	Four ATP molecules are produced/generated/made in the energy pay-off stage **OR** there is a net gain of two ATP	
	5	Pyruvate is the end product **(Any 3)**	
	6	If oxygen is available/in aerobic conditions pyruvate progresses to the citric acid cycle	
	7	Pyruvate is converted/broken down to an acetyl group	
	9	The acetyl group combines with coenzyme A	
	8	Acetyl (coenzyme A) combines with oxaloacetate to form citrate	
	10	The citric acid cycle is enzyme controlled/involves dehydrogenases	
	11	ATP is generated/synthesised/produced/released in the citric acid cycle	
	12	Carbon dioxide is released from the citric acid cycle	
	13	Oxaloacetate is regenerated	
	14	NAD/NADH transports electrons **and** hydrogen ions to the electron transport chain **(Any 6)**	
39	1	The electron transport chain takes place on the inner membrane of the mitochondria	8
	2	The electron transport chain is a collection of proteins attached to the inner membrane of a mitochondrion	
	3	NADH releases the electrons to the electron transport chain on the inner mitochondrial membrane	
	4	The electrons pass down the chain of electron acceptors/proteins, releasing their energy	
	5	The energy is used to pump hydrogen ions (H^+) across the inner mitochondrial membrane	
	6	The return flow of the hydrogen ions (H^+) back into the matrix drives the enzyme ATP synthase	
	7	This results in the synthesis of ATP from ADP + Pi	
	8	This stage produces most of the ATP generated by cellular respiration	
	9	Oxygen is the final hydrogen ion and electron acceptor	
	10	Water is formed **(Any 8)**	

11 Metabolic rate

What you need to know about measuring metabolic rates

The **metabolic rates** of different organisms at rest can be compared through the measurement of oxygen consumption, carbon dioxide production and heat production.

Metabolic rate can be measured using respirometers, oxygen probes, carbon dioxide probes and calorimeters.

Key diagram

A calorimeter designed to measure metabolic rate in a human subject

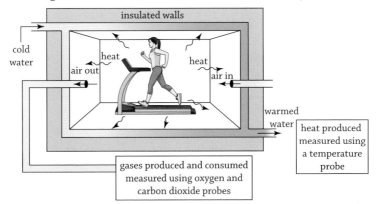

Technique

Measuring metabolic rate is a technique which you need to be familiar with for your exam.

Metabolic rate is often investigated directly by measuring the heat produced by living material. It can also be investigated indirectly by measuring the rate of respiration in living material assuming that this is proportional to metabolic rate. You should look again at the Key diagram above.

C-type questions

1. State what is meant by an organism's metabolic rate.   1
2. Give **two** instruments that can be used to compare metabolic rates in different organisms.   2

A-type question

3. Describe **two** measurements which can be used in determining the metabolic rate of an organism.   2

➡ **Model answers and commentary can be found on page 88.**

What you need to know about delivery of oxygen to cells

Organisms with high metabolic rates require more efficient delivery of oxygen to cells.

Birds and mammals have higher metabolic rates than reptiles and amphibians, which in turn have higher metabolic rates than fish.

Birds and mammals have a **complete double circulatory system** consisting of two atria and two ventricles.

Amphibians and most reptiles have an **incomplete double circulatory system** consisting of two atria and one ventricle.

Fish have a **single circulatory system** consisting of one atrium and one ventricle.

Complete double circulatory systems enable higher metabolic rates to be maintained.

In complete double circulatory systems, there is no mixing of oxygenated and deoxygenated blood and the oxygenated blood can be pumped out at a higher pressure, which enables more efficient oxygen delivery to cells.

Key diagram

Comparison of circulation in fish, amphibians and most reptiles, and birds and mammals
(a)   Diagram

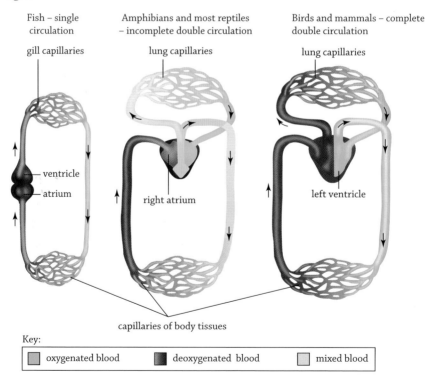

Fish – single circulation

gill capillaries

ventricle
atrium

Amphibians and most reptiles – incomplete double circulation

lung capillaries

right atrium

Birds and mammals – complete double circulation

lung capillaries

left ventricle

capillaries of body tissues

Key:
☐ oxygenated blood   ☐ deoxygenated blood   ☐ mixed blood

(b) Table

Vertebrate group	Blood pressure level	Oxygenation of blood to tissues
Fish	Low	Fully oxygenated
Amphibians and most reptiles	High	Partly oxygenated
Birds and mammals	High	Fully oxygenated

C-type questions

4. Name the type of circulation in amphibians.   1

5. Name the type of circulation in birds and mammals.   1

A-type questions

6. Explain why the structure of a mammalian heart makes it more efficient than the heart of a fish.   2

7. Explain why organisms with a high metabolic rate require an efficient delivery of oxygen to their cells.   2

8. Explain how the heart of a mammal is better adapted than the heart of an amphibian to allow high metabolic rates in the cells that it supplies with blood.   2

9. Explain why the circulatory system of mammals needs to be more efficient than those of the other vertebrates shown.   2

 Many candidates find relating heart structure to efficiency of oxygen delivery and metabolic rate difficult – remember to relate heart structure to blood pressure and levels of blood oxygenation.

Extended response questions

10. Describe the arrangement of heart chambers in birds and amphibians and relate this to their metabolic rates.   4

11. Compare and contrast the heart structure and circulation of fish, amphibians and mammals.   9

Model answers and commentary

Question		Model answer	Marks	Commentary with hints and tips
1		The energy used in a given period of time	1	Make into a flash card. Remember that whenever you are asked about 'rate' you must mention 'in a given period of time' such as per minute or per hour.
2		Respirometer; calorimeter	2	Only the names are needed.
3		Oxygen consumed in a given period of time; Carbon dioxide produced in a given period of time; The energy released as heat in a given period of time **(Any 2)**	2	Rate = in a given period of time.
4		Incomplete double circulatory system	1	A three-chambered heart made up of a right and left atrium and one ventricle in which oxygenated and deoxygenated blood mix.
5		Complete double circulatory system	1	A four-chambered heart made up of two atria and two ventricles in which oxygenated and deoxygenated blood do not mix.
6		Mammals have a four-chambered heart/two ventricles; delivers oxygenated blood at higher pressure to tissues	2	In fish, the loss of pressure is a problem.
7		More oxygen so more aerobic respiration; More ATP needed to maintain high metabolic rate	2	Organisms with high metabolic rates require more ATP, so more oxygen is needed for more respiration.
8		In mammals, fully oxygenated blood can be delivered to the muscles/tissues because heart has two ventricles; In amphibians, partly/incompletely oxygenated blood is delivered to the muscles/tissues because heart has only one ventricle where blood can mix	2	In amphibians the blood from both atria is passed into the one ventricle, which means that the oxygenated and deoxygenated blood mix.

9		Mammals have higher metabolic rates; They require more efficient delivery of oxygen to cells	2	Organisms with high metabolic rates require more ATP, so more oxygen is needed for more respiration.
10	1	Amphibian heart has two atria and one ventricle	4	
	2	Bird heart has two atria and two ventricles		
	3	Birds have a high metabolic rate **OR** Amphibians have a low metabolic rate		
	4	No mixing of oxygenated and deoxygenated blood in bird heart **OR** Mixing of oxygenated and deoxygenated blood in amphibians		
	5	More efficient oxygen delivery to bird cells/muscles/tissues in birds **OR** Less efficient oxygen delivery to cells/muscles/tissues in amphibians **(Any 4)**		
11		Fish:	9	
	1	Single circulatory system/heart to gills to body to heart		
	2	Heart with only two chambers/atria and a ventricle		
	3	Loss of pressure a problem		
		Amphibian:		
	4	Incomplete double circulatory system		
	5	Heart with three chambers/right and left atrium and one ventricle		
	6	Pressure maintained		
	7	Tissue blood is incompletely oxygenated		
		Mammal:		
	8	Complete double circulatory system		
	9	Heart has four chambers/two atria and two ventricles		
	10	Pressure is maintained		
	11	Tissue blood is completely oxygenated **(Any 9)**		

CHAPTER 12

Metabolism in conformers and regulators

What you need to know about metabolic rate and abiotic factors

The ability of an organism to maintain its metabolic rate is affected by external abiotic factors.

Abiotic factors include temperature, salinity and pH.

Key diagram

Abiotic factors which can affect metabolic rate

temperature

pH of soil or water

salinity

C-type questions

1. Name **two** abiotic factors that can affect the ability of an organism to maintain its metabolic rate.　　　　　　　　　　　　　　　　　2

2. Name **one** external abiotic factor, other than temperature, which can affect the ability of an organism to maintain its metabolic rate.　　　　1

A-type question

3. Explain how a named abiotic factor can affect the ability of an organism to maintain its metabolic rate.　　　　　　　　　　　　　　　　　2

➡ **Model answers and commentary can be found on page 95.**

What you need to know about conformers

Conformers' internal environment is dependent upon the external environment.

Conformers use behavioural responses to maintain optimum metabolic rate.

Conformers have low metabolic costs and a narrow range of **ecological niches**.

Behavioural responses by conformers allow them to tolerate variation in their external environment to maintain optimum metabolic rate.

Key diagram

Conformer using behaviour to collect heat from its environment to raise its metabolic rate to optimum

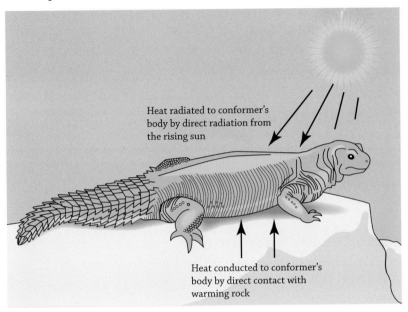

Heat radiated to conformer's body by direct radiation from the rising sun

Heat conducted to conformer's body by direct contact with warming rock

C-type questions

4. Give the term used to describe organisms whose internal environment is dependent on their external environment.   1

5. Name **one** type of response that can help to maintain a conformer's optimum metabolic rate.   1

A-type questions

6. Explain why conformers have low metabolic costs.   2

7. Explain why conformers usually have a narrow ecological niche.   2

➡ **Model answers and commentary can be found on page 95.**

What you need to know about regulators

Regulators maintain their internal environment regardless of external environment.

Regulators use metabolism to control their internal environment, which increases the range of possible ecological niches.

Regulators require energy to achieve **homeostasis**. This increases their metabolic costs.

Key diagram

Example of homeostasis – body temperature control in a mammal compared with a lizard

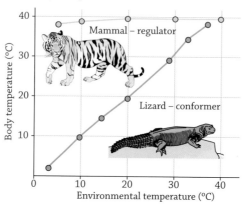

C-type questions

8. Give the meaning of the term 'homeostasis'. 1
9. Give the term used to describe organisms with the ability to control their internal environment by metabolic means. 1

A-type questions

10. Explain why regulators can occupy a wide range of ecological niches. 2
11. Explain why regulators have high metabolic cost. 1

Model answers and commentary can be found on page 95.

What you need to know about thermoregulation

Thermoregulation is the control of body temperature by **negative feedback**.

The **hypothalamus** is the temperature-monitoring centre in the brain of a mammal.

Information is communicated by electrical impulses through nerves to the effectors, which bring about corrective responses to return temperature to normal.

Corrective responses to an increase in body temperature include increased sweating, vasodilation of blood vessels and decreased metabolic rate.

Body heat is used to evaporate water in the sweat, cooling the skin.

Vasodilation results in increased blood flow to the skin, which increases heat loss.

A reduction in the metabolic rate results in less heat being produced.

Corrective responses to a decrease in body temperature include shivering, vasoconstriction of blood vessels, hair erector muscles contracting and increased metabolic rate.

Muscle contraction due to shivering generates heat.

Vasoconstriction results in decreased blood flow to the skin, which decreases heat loss.

Hair erector muscles contract, raising hairs and trapping a layer of insulating air.

Increasing the metabolic rate produces more heat.

Key diagram

Thermoregulation by negative feedback control in mammals, including the roles of the hypothalamus, nerves and effectors

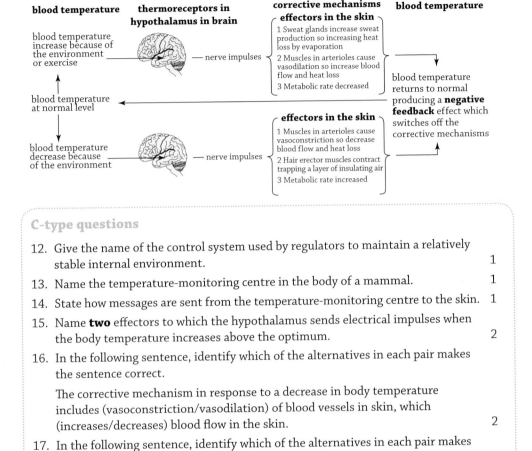

C-type questions

12. Give the name of the control system used by regulators to maintain a relatively stable internal environment.   1

13. Name the temperature-monitoring centre in the body of a mammal.   1

14. State how messages are sent from the temperature-monitoring centre to the skin.   1

15. Name **two** effectors to which the hypothalamus sends electrical impulses when the body temperature increases above the optimum.   2

16. In the following sentence, identify which of the alternatives in each pair makes the sentence correct.

    The corrective mechanism in response to a decrease in body temperature includes (vasoconstriction/vasodilation) of blood vessels in skin, which (increases/decreases) blood flow in the skin.   2

17. In the following sentence, identify which of the alternatives in each pair makes the sentence correct.

    The corrective mechanism in response to an increase in body temperature includes (vasoconstriction/vasodilation) of blood vessels in skin, which (increases/decreases) blood flow in the skin.   2

A-type questions

18. Describe the response of a mammal to an increase in body temperature.   2

19. Explain the advantage to a mammal of the vasoconstriction of its blood vessels when skin temperature drops from 20°C to 5°C.   2

20. Describe the role of negative feedback control in thermoregulation.   2

➡ **Model answers and commentary can be found on page 96.**

What you need to know about thermoregulation and metabolism

Regulating temperature (thermoregulation) is important for optimal enzyme-controlled reaction rates and high diffusion rates for the maintenance of metabolism.

Key diagram

Diagram of a cell showing the processes for which temperature is important

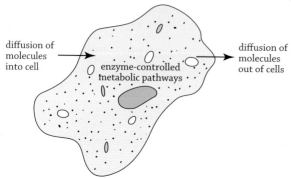

diffusion of molecules into cell

enzyme-controlled metabolic pathways

diffusion of molecules out of cells

Many candidates have difficulty in explaining why animals regulate body temperature – remember enzymes and diffusion.

C-type questions

21. Give **one** reason why it is important for mammals to regulate their body temperature. 1
22. Give **two** reasons why body temperature in mammals is important to metabolic processes.   2

A-type questions

23. Explain why regulating body temperature is important to the metabolism of mammals.   2
24. Explain the importance of thermoregulation to mammals.   2

Extended response questions

25. Describe and compare metabolism in conformers and regulators.   5
26. Explain the principle of negative feedback in relation to the control of body temperature.   5
27. Give an account of metabolism in conformers in relation to their ecological niches. 7
28. Give an account of the mechanisms of thermoregulation in mammals in response to a decrease in body temperature.   9

Model answers and commentary

Question		Model answer	Marks	Commentary with hints and tips
1		Temperature; salinity; pH **(Any 2)**	2	Abiotic factors are non-living variables that can influence where organisms can live.
2		pH **OR** salinity	1	
3		Temperature/pH; affects enzyme activity **OR** Salinity; affects osmoregulation and more metabolic activity needed to maintain osmotic balance	2	Your explanation must link the named abiotic factor to its effect on metabolism.
4		Conformers	1	Conformers cannot alter their metabolic rate using physiological means, but as a result their metabolic costs can be low.
5		Behavioural	1	It is good to be able to give a few examples, such as some reptiles basking on rocks to increase their body temperature.
6		They do not use physiological mechanisms to control/alter their metabolic rate; Physiological mechanisms require energy	2	Low metabolic cost means that they have a lower energy requirement.
7		Conformers lack the ability to tolerate environmental change should it occur **OR** They need a stable environment; They cannot use physiological mechanisms to control/alter their metabolic rate **OR** Conformers' internal environments are directly dependent upon their external environment	2	Conformers live in stable environments such as the ocean depths.
8		The maintenance of steady body conditions	1	This is another need-to-know term – one for the flash cards.
9		Regulators	1	Animals that can adjust their metabolic rate to maintain a steady internal state.

10		Because they can maintain/ control their internal environment in a wide range of conditions; using their metabolism	2	Remember to link regulators with metabolism.
11		They use energy to maintain homeostasis	1	Remember to link regulators with the use of energy.
12		Negative feedback	1	The control mechanism by which homeostasis is achieved.
13		Hypothalamus	1	Located in the brain, it contains thermoreceptors which detect changes in blood temperature.
14		Electrical impulses	1	The hypothalamus sends out electrical impulses to effectors in the skin and body muscles.
15		Sweat glands; skin arterioles; hair erector muscles **(Any 2)**	2	Responses to an increase in temperature include: sweating increased, vasodilation, hairs lowered.
16		Vasoconstriction; decreases	2	This results in decreased heat loss by radiation.
17		Vasodilation; increases	2	This results in increased heat loss by radiation.
18		Increased sweat production; vasodilation of blood vessels/ arterioles; hair erector muscles relaxed **(Any 2)**	2	A description rather than an explanation is required for this question. Take care **not** to say vasodilation of capillaries.
19		(Vasoconstriction) reduces the volume of blood flow to skin surface; Reduces heat loss by radiation	2	An explanation needs detail of the increased blood volume and the cooling effect due to heat loss by radiation.
20		Ensures that when the body temperature returns to normal/ optimum; the corrective mechanisms are switched off	2	'Corrective mechanisms switched off' is needed for the second mark.
21		Enzymes have an optimum temperature at which they are most active **OR** Diffusion rates are affected by temperature	1	No explanation is needed for this answer.
22		Enzymes have an optimum temperature at which they are most active; Diffusion rates are affected by temperature	2	Two reasons are needed for this answer.

23		Mammals maintain the optimum temperature for enzyme activity so can maintain a high metabolic rate; Rates of diffusion of substances such as oxygen and carbon dioxide are faster, which helps to maintain a high metabolic rate	2	An explanation linked to metabolism and metabolic rate is needed this time.	
24		Mammals can maintain the optimum temperature for enzyme activity so can maintain a high metabolic rate; Rates of diffusion of substances such as oxygen and carbon dioxide are faster, which helps to maintain a high metabolic rate	2	You need to be able to recognise this rewording of Question 21. An explanation linked to metabolism and metabolic rate is needed this time.	
25	1	Conformers' metabolism/metabolic rate/internal environment is dependent on/affected by surroundings/external environment/ external factors/external variables			5
	2	Conformers use behaviour to maintain optimum metabolic rate			
	3	Regulators can maintain/control/regulate their metabolism/ metabolic rate/internal environment regardless of external conditions			
	4	Regulators require energy for homeostasis/negative feedback			
	5	Conformers have narrower (ecological) niches (or converse)			
	6	Conformers have lower metabolic costs/rates of metabolism (or converse) **(Any 5)**			
26	1	A mechanism for achieving homeostasis			5
	2	Maintains a constant internal environment/keeps internal environment steady			
	3	Any change away from the norm/set point/optimum is detected by a receptor/hypothalamus			
	4	Switches on corrective mechanisms to reverse the change using effectors			
	5	Messages sent by nerve impulses			
	6	Once corrective mechanism has restored conditions back to the norm/set point/optimum, the strength of the response decreases/ the corrective mechanism is switched off **(Any 5)**			

27	1	Ability of an organism to maintain its metabolic rate is affected by external abiotic factors such as temperature/salinity/pH	7
	2	Conformers' internal environment is dependent upon its external environment	
	3	Conformers cannot alter their metabolic rate using physiological means	
	4	Conformers usually have a narrow ecological niche/limited range	
	5	Conformers lack the ability to tolerate change should it occur	
	6	Conformers live in stable environments	
	7	Conformers do not use energy-requiring physiological mechanisms to alter their metabolic rate	
	8	Conformers have low metabolic energy costs	
	9	Many conformers manage to maintain their optimum metabolic rate by employing certain behavioural responses **(Any 7)**	
28	1	Temperature-monitoring centre/thermoreceptors are located in the hypothalamus **OR** Information about temperature detected/received by hypothalamus	9
	2	Mammals obtain most of their body heat from respiration/metabolism/chemical reactions	
	3	Nerve message/communication/impulse sent to skin/effectors	
	4	Vasoconstriction/narrowing of arterioles to skin in response to decreased temperature	
	5	Less/decreased blood flow to skin/extremities	
	6	Decreased/less heat radiated from skin/extremities	
	7	Decrease in temperature causes hair erector muscles to raise/erect hair	
	8	Traps warm air **OR** Forms insulating layer	
	9	Decrease in temperature causes muscle contraction/shivering, which generates heat/raises body temperature	
	10	Metabolic rate increases	
	11	Temperature regulation involves/is an example of negative feedback **(Any 9)**	

13 Metabolism and adverse conditions

Many environments vary beyond the tolerable limits for normal metabolic activity for any particular organism. Some animals have adapted to survive these adverse conditions while others avoid them.

What you need to know about surviving adverse conditions

Dormancy is part of some organisms' life-cycle to allow survival during a period when the costs of continued normal metabolic activity would be too high.

The metabolic rate can be reduced during dormancy to save energy.

During dormancy there is a decrease in metabolic rate, heart rate, breathing rate and body temperature.

Dormancy can be **predictive** or **consequential**.

Predictive dormancy occurs before the onset of adverse conditions.

Consequential dormancy occurs after the onset of adverse conditions.

Some mammals survive during winter/low temperatures by **hibernating**.

Aestivation allows survival in periods of high temperature or drought.

Daily **torpor** is a period of reduced activity in some animals with high metabolic rates.

Key diagram

Summary of dormancy behaviours in animals

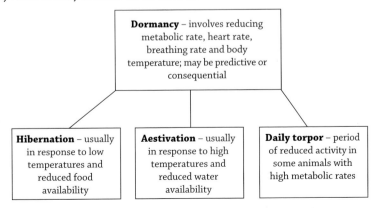

C-type questions

1. Give the meaning of the term 'dormancy'. — 1

2. Give **two** examples of dormancy. — 2

3. Give the term for the type of dormancy that occurs before the onset of adverse conditions. — 1

4. Give the term for the type of dormancy that occurs after the onset of adverse conditions. — 1

5. Name the type of dormancy typical of organisms living in unpredictable environments. — 1

6. Some species of small mammal with high metabolic rates enter a state of reduced activity each day to survive adverse conditions.

   Give the term used to describe this state. — 1

7. Give **two** factors that can contribute to the metabolic energy crisis faced by mammals that hibernate. — 2

8. Give **one** behavioural adaptation of animals with high metabolic rates to allow survival in adverse conditions. — 1

9. Name the form of dormancy that allows some animals to survive in periods of high temperature or drought in the summer. — 1

10. Give the meaning of the term 'daily torpor'. — 1

A-type questions

11. Describe the difference between consequential and predictive dormancy. — 2

12. Explain the benefit to some animals of being able to undergo daily torpor. — 2

13. Explain the survival role of aestivation. — 2

14. Explain the survival role of hibernation in some mammals. — 2

➡ **Model answers and commentary can be found on page 101.**

What you need to know about avoiding adverse conditions

Some organisms avoid adverse conditions by undertaking **migration**.

Migration avoids metabolic adversity by expending energy to relocate to a more suitable environment.

Migratory behaviour can be **innate** and **learned**.

Specialised techniques are used to study long-distance migration. Examples of specialist techniques include satellite tracking and leg rings.

Key diagram

Flocks of birds avoiding metabolic adversity by migrating: **(a)** black-tailed godwits migrate innately but **(b)** sandhill cranes learn the behaviour from their parents.

(a)    **(b)**

C-type questions

15. Many species of bird avoid metabolic adversity by relocating to a more suitable environment.
    Name this type of behaviour.   1
16. Give **one** example of how organisms can avoid adverse conditions.   1
17. Give the meanings of innate and learned behaviour in bird migration.   2
18. Give **one** disadvantage of migration.   1
19. Name **two** methods of tracking migratory animals.   2

A-type questions

20. Explain the benefit to some animals of being able to migrate.   1
21. Describe a technique which could be used to track species which undertake
    long-distance migration.   1

Extended response questions

22. Write notes on adaptations shown by organisms to survive adverse conditions.   5
23. Give an account of the adaptations of organisms to surviving and avoiding
    adverse environmental conditions.   8

Model answers and commentary

Question	Model answer	Marks	Commentary with hints and tips
1	Period of reduced metabolic activity to survive adverse conditions **OR** Period of reduced metabolic rate to allow survival when the energy required for normal activity would be too high	1	Use this answer to make a flash card. Dormancy is part of an organism's life-cycle and is the stage associated with resisting or tolerating periods of environmental adversity.
2	Hibernation; aestivation; daily torpor **(Any 2)**	2	Make flash cards with definitions for each of these.

3		Predictive dormancy	1	Predictive dormancy is when an organism becomes dormant before onset of the adverse conditions. It is usually genetically programmed and is triggered by seasonal environmental cues/stimuli.
4		Consequential dormancy	1	Consequential dormancy is when an organism becomes dormant after the onset of adverse conditions. It is a typical response of organisms living in unpredictable environments.
5		Consequential dormancy	1	Unpredictable environments do not provide the environmental trigger stimuli, such as decreasing photoperiod, associated with predictive dormancy.
6		Daily torpor	1	Torpor is a period of reduced activity in organisms with high metabolic rates. It increases an organism's chances of survival by reducing the energy required to maintain a high metabolic rate. It is similar to short-term hibernation.
7		Low temperatures; lack of food	2	These adverse conditions result in an organism not being able to meet its metabolic costs. During hibernation, metabolic rate is reduced, resulting in a decrease in body temperature, heart rate and breathing rate.
8		Hibernation; aestivation; daily torpor **(Any 1)**	1	Similar to Question 2 but reworded to test that you are aware that these are behavioural adaptations.
9		Aestivation	1	Usually involves burrowing into the ground, where the temperature stays cool, and reducing metabolic activity in a similar manner to hibernation.
10		A period of reduced activity/ metabolism each day to survive adverse conditions	1	Torpor increases an organism's chances of survival by reducing the energy required to maintain a high metabolic rate for part of every 24-hour cycle.

11		Predictive dormancy is when an organism becomes dormant before onset of the adverse conditions; Consequential dormancy is when an organism becomes dormant after the onset of adverse conditions	2	Predictive dormancy is usually genetically programmed and is triggered by seasonal environmental cues/stimuli. Consequential dormancy is a typical response of organisms living in unpredictable environments.
12		It increases an organism's chances of survival; By reducing the energy required to maintain a high metabolic rate	2	An explanation is needed rather than just a description. It decreases the rate of energy consumption when the animal would be unable to obtain enough food.
13		Aestivation allows survival in periods of high temperature/ drought; When there would be a lack of food/water	2	Link the high temperatures to the drought conditions and how this could lead to the death of the organism.
14		Hibernation allows survival in periods of low temperatures/ lack of food; When organism would not able to meet its metabolic costs/demand	2	During hibernation, metabolic rate is reduced, resulting in a decrease in body temperature, heart rate and breathing rate.
15		Migratory behaviour/migration	1	Remember that to 'avoid' adverse conditions the organism needs to move away. Migration is a relatively long-distance movement of individuals, which usually takes place on a seasonal basis.
16		Migration	1	Migration enables animals to avoid metabolic adversity brought about by lack of food and low temperatures.
17		Innate behaviour is unlearned/ instinctive behaviour; Learned behaviour is acquired by experience	2	Some birds (e.g. black-tailed godwit) have an innate homing ability, while others follow their parents. Cranes learn their migration route from older cranes and get better at it with age.
18		Migration expends/needs energy/has a high metabolic cost **OR** Animals expend energy to relocate to a more suitable environment	1	Relocating to a favourable environment increases survival chances. Net energy gain.

19		Ringing; tagging; transmitters **(Any 2)**	2	Remember '**RR**': **R**ing and **R**ecapture.
20		Adverse conditions/metabolic adversity avoided by migration **OR** Migration enables animals to avoid metabolic adversity brought about by lack of food and low temperatures	1	Migration enables animals to avoid metabolic adversity brought about by lack of food and low temperatures.
21		Ringing **AND** recapture; Tags using GPS; Transmitters **AND** receivers/satellites **(Any 1)**	1	It is not enough to just name a technique – a description of use is also required.

22	1	Some environments vary beyond tolerable limits	5
	2	Extremes of conditions do not allow the normal metabolism of the organisms present	
	3	Variation in conditions can be cyclical or unpredictable	
	4	Metabolic rate can be reduced when conditions would make the cost of metabolic activity too high	
	5	Dormancy may be predictive **OR** consequential	
	6	Example from hibernation **OR** aestivation	
	7	Daily torpor is a period of reduced activity **(Any 5)**	
23	1	Animals survive adverse conditions/metabolic adversity by dormancy	8
	2	One type named (hibernation, aestivation or daily torpor)	
	3	Another named type	
	4	Dormancy is where metabolic rate/heart rate/breathing rate/body temperature is reduced	
	5	Dormancy/hibernation/aestivation/daily torpor conserves/saves energy	
	6	Predictive dormancy/hibernation occurs before the onset of adverse conditions/correct description of adverse conditions	
	7	Consequential dormancy/hibernation/aestivation occurs after the onset of adverse conditions/correct description of adverse conditions	
	8	Hibernation occurs in times of low temperatures/winter/cold conditions	
	9	Aestivation occurs in times of drought/high temperature	
	10	Daily torpor occurs in animals with high metabolic rates **(Any 6)**	
	11	Adverse conditions/metabolic adversity avoided by migration	
	12	Migration expends/needs energy/has a high metabolic cost	
	13	Migration is innate and/or learned (both terms required) **(Any 2)**	

CHAPTER 14
Environmental control of metabolism in microorganisms

What you need to know about microorganisms, growth media and control of environmental factors

Microorganisms are archaea, bacteria and some species of eukaryotes.

Microorganisms use a wide variety of substrates for metabolism and produce a range of products from their metabolic pathways.

Microorganisms are used in industry and pharmaceutics because of their adaptability, ease of cultivation and speed of growth.

Many microorganisms produce all the complex molecules required for biosynthesis, for example amino acids, vitamins and fatty acids.

Other microorganisms require complex molecules to be supplied in their **growth media**.

When culturing microorganisms, their growth media requires raw materials for biosynthesis as well as an energy source.

An energy source is derived either from chemical substrates or from light in photosynthetic microorganisms.

Culture conditions include **sterility** and the control of temperature, pH and oxygen levels.

Sterile conditions in fermenters reduce competition with desired microorganisms for nutrients and reduce the risk of spoilage of the product.

Key diagram

Features of a fermenter which provide optimum conditions for the growth of microorganisms

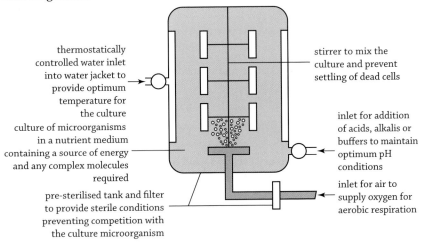

thermostatically controlled water inlet into water jacket to provide optimum temperature for the culture

culture of microorganisms in a nutrient medium containing a source of energy and any complex molecules required

pre-sterilised tank and filter to provide sterile conditions preventing competition with the culture microorganism

stirrer to mix the culture and prevent settling of dead cells

inlet for addition of acids, alkalis or buffers to maintain optimum pH conditions

inlet for air to supply oxygen for aerobic respiration

Many candidates have trouble naming a complex compound used in growth media – remember to use amino acids, vitamins and fatty acids as your examples.

Many candidates have trouble identifying the process for which microorganisms need amino acids in the context of cell culture in a fermenter – remember to make the link to protein synthesis.

➡ See 'What you need to know about translation' in Chapter 3.

Many candidates have trouble explaining the importance of preventing contamination of a cell culture in a fermenter – remember that contaminating microorganisms are competition for the desired microorganisms.

C-type questions

1. Give **two** features of microorganisms that make them useful for a variety of research and industrial uses. 2

2. State **two** factors that are controlled to provide optimum culture conditions for growth of microorganisms in a fermenter. 2

3. In culturing bacteria it is important to control the pH of the culture medium.
   Give **one** method of controlling the pH of a culture medium. 1

4. State why some fermenters have steam pumped through them before adding the culture of microorganisms. 1

5. Give **two** complex compounds that are sometimes added to culture media to enable certain microorganisms to grow. 2

A-type questions

6. In culturing bacteria it is important to control the pH of the culture medium.
   Explain why the pH of a culture medium should be controlled. 2

7. Explain why sterile conditions must be maintained in the culture of a microorganism. 2

8. Explain why temperature, pH and oxygen levels must be monitored and controlled during the culture of microorganisms. 2

➡ Model answers and commentary can be found on page 109.

What you need to know about phases of growth and changes in culture conditions

Phases of growth in microorganisms include lag, log/exponential, stationary and death.

The **lag phase** is where enzymes are induced to metabolise substrates.

The **log/exponential phase** contains the most rapid growth of microorganisms due to plentiful nutrients.

The **stationary phase** occurs due to the nutrients in the culture media becoming depleted and the production of toxic metabolites.

In the stationary phase, **secondary metabolites** such as antibiotics are also produced.

In the wild, secondary metabolites confer an ecological advantage by allowing the microorganisms which produce them to outcompete other microorganisms.

The **death phase** occurs due to the accumulation of toxic metabolites or the lack of nutrients in the culture.

Semi-logarithmic scales are used in producing or interpreting growth curves of microorganisms.

Viable cell counts involve counting only the living microorganisms whereas total cell counts involve counting viable and dead cells.

Only viable cell counts show a death phase where cell numbers are decreasing.

Key diagram

Graph showing the phases of the growth curve of microorganisms plotted on semi-logarithmic scales

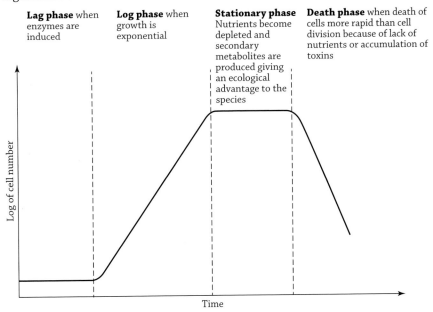

Lag phase when enzymes are induced

Log phase when growth is exponential

Stationary phase Nutrients become depleted and secondary metabolites are produced giving an ecological advantage to the species

Death phase when death of cells more rapid than cell division because of lack of nutrients or accumulation of toxins

Log of cell number

Time

Many candidates find difficulty in identifying the phase of growth where secondary metabolites are produced – remember 'SS': Stationary and Secondary.

Many candidates find difficulty in suggesting an ecological advantage gained by producing a secondary metabolite – make the link with competition with other bacteria which are killed by antibiotics.

Many candidates find difficulty in explaining how a growth curve shows viable cell count – remember that total cell counts include both living and dead cells but viable counts include only live cells.

C-type question

9. Secondary metabolism can confer an ecological advantage to a microorganism by producing substances not associated with growth.

    Give **one** example of a secondary metabolite.   1

A-type questions

10. Describe what happens during the lag phase of microbial growth.   2
11. Describe what happens during the log phase of microbial growth.   2
12. Explain why the number of microorganisms remains constant during the stationary phase of microorganisms grown in culture.   2
13. Explain what is meant by the total cell count, which is used to estimate the populations of microorganisms.   1
14. Explain why substances produced during secondary metabolism might give an ecological advantage to a microorganism.   2
15. Describe the role of secondary metabolites and enzyme induction in the growth of microorganisms.   2

Extended response questions

16. Give an account of the phases of growth of microorganisms cultured in a fermenter.   7
17. Give an account of the different culture conditions required for the growth of microorganisms.   9

Model answers and commentary

Question		Model answer	Marks	Commentary with hints and tips
1		They use a wide range of substrates for metabolism; Produce a wide range of products from their metabolic pathways; Ease of cultivation; Speed of growth; Cheap to use **(Any 2)**	2	Use of the term 'features of microorganisms' sometimes confuses candidates. These are features/characteristics that make them useful. Learn at least two of them.
2		Sterility; temperature; oxygen levels; pH; suitable media/nutrients **(Any 2)**	2	Remember '**STOP**': **S**terility, **T**emperature, **O**xygen, **p**H.
3		Buffers **OR** the addition of acid **OR** alkali	1	A buffer solution is one which resists changes in pH when small quantities of an acid or an alkali are added to it. Needed here to maintain optimum pH for enzyme activity.
4		Sterility/sterilise the fermenter/ container/to kill other/unwanted microorganisms **OR** to prevent contamination from unwanted microorganisms	1	Other microorganisms would use/compete for the substrate and not produce the desired product. They may reduce productivity/ growth/yield of the culture.
5		Vitamins; fatty acids; amino acids **(Any 2)**	2	Complex media are rich in nutrients; they contain water-soluble extracts of plant or animal tissue. They also contain a carbon source such as glucose; water; various salts; a source of amino acids and nitrogen (e.g. **beef extract**, yeast extract). Notice that complex compounds = vitamins/ fatty acids/amino acids and complex ingredients refer to the beef extract or yeast extract.

6		pH is a factor which affects rates of enzyme activity; In fermenters buffers can be added to maintain optimum pH conditions for enzyme activity	2	Sensors monitor the culture conditions and maintain the factors affecting growth at their optimum level. Needed to maintain the optimum pH for microbial enzyme activity.
7		To prevent contamination from unwanted microorganisms; Other microorganisms would use/compete for the substrate and not produce the desired product **OR** They may reduce productivity/growth/yield of the culture **OR** They may cause disease/health risks/harm humans	2	'Explain' questions need you to go further. For example, sterility prevents contamination of unwanted microorganisms, but why is that important?
8		Optimum temperature and pH for microorganism enzymes; Oxygen for aerobic respiration of microorganisms	2	Need to create optimum conditions for enzyme activity and aerobic respiration.
9		Antibiotic **OR** pigment	1	Antibiotics inhibit the growth of contaminating bacteria and so reduces competition for the limited resources.
10		The microorganisms induce the production/synthesis of required enzymes; That metabolise the substrates **OR** That are needed for the primary metabolism/log phase	2	No cell division during the lag phase as the enzymes needed for the primary metabolism in the log phase are being synthesised.
11		Substrate is broken down to obtain energy/ATP; Population doubles with each cell division **OR** Produces primary metabolites used for the biosynthesis of substances such as proteins/nucleic acids	2	The log or exponential phase of growth.
12		Cell division equals number of cells dying; Because resources/nutrients limiting/toxic metabolites building up	2	Although cells still dividing, some are dying.
13		The total number of cells in a culture including viable/live cells **and** dead cells	1	Remember viable means still alive.

14		Results in the production of secondary metabolites such as antibiotics; Inhibit the growth of other bacteria **AND** so reduce competition for the available/limited resources	2	Remember '**SS**': **S**econdary metabolites produced during the **S**tationary phase. In their environment, the microorganisms producing the antibiotics reduce competition for the remaining resources. In the fermenter, the population size must be controlled very carefully to ensure that the maximum yield of antibiotics is obtained before the cells die.
15		The enzymes are required for the primary metabolism/log phase/to metabolise the substrates; The secondary metabolites/antibiotics inhibit the growth of other bacteria/reduce competition for the available/limited resources	2	This needs a real understanding of the events that occur during the phases of growth of microorganisms cultured in a fermenter. Practise drawing the graph, labelling the phases and adding the additional information.
16	1	Growth is recorded by measuring the increase in cell number in a given period of time	7	
	2	The time it takes for a unicellular organism to divide into two is called the doubling **OR** generation time		
	3	The lag phase of growth is where the microorganisms induce the production/synthesis of enzymes that metabolise the substrates		
	4	No cell division occurs at this stage		
	5	The log/exponential phase of growth is where the population doubles with each cell division		
	6	The stationary phase is where the culture medium becomes depleted/nutrients **OR** oxygen start to run out		
	7	The stationary phase is reached when the rate of production of new cells is equal to the death rate of the older cells		
	8	The death phase occurs due to lack of substrate/toxic accumulation of metabolites		
	9	More cells die than are being produced **(Any 7)**		

17	1	Require an energy source	9
	2	Energy is derived from substrates such as carbohydrates/from light in the case of photosynthetic microorganisms	
	3	Supply of raw materials for the biosynthesis of proteins/nucleic acids	
	4	Many microorganisms only require simple chemical compounds in growth media	
	5	Some microorganisms require specific complex compounds in growth media	
	6	Example of complex compound, e.g. vitamins/fatty acids/amino acids	
	7	Growth media can contain complex ingredients such as beef extract	
	8	Sterility to eliminate any effects of contaminating microorganisms	
	9	Control of temperature/pH/oxygen	
	10	Control of oxygen levels by aeration	
	11	Control of pH by buffers **OR** the addition of acid or alkali **(Any 9)**	

Genetic control of metabolism in microorganisms

What you need to know about improving wild strains of microorganism

Wild strains of microorganisms can be improved by **mutagenesis** or **recombinant DNA technology**.

Exposure to ultraviolet (UV) light and other forms of radiation or mutagenic chemicals results in mutations, some of which may produce an improved strain of microorganism.

Key diagram

Wild strain of microorganism improved by mutagenesis and recombinant DNA technology

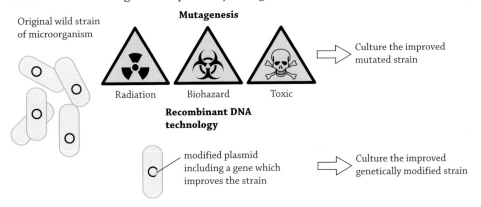

C-type questions

1. State **two** methods by which wild strains of microorganisms could be improved.   2
2. Give **two** mutagenic agents that can be used to increase the rate of mutation in microorganisms.   2

A-type question

3. Explain what is meant by the term 'mutagenesis'.   1

➡ **Model answers and commentary can be found on page 116.**

What you need to know about recombinant DNA technology

Recombinant DNA technology involves the use of recombinant plasmids and **artificial chromosomes** as vectors.

A vector is a DNA molecule used to carry foreign genetic information into another cell, and both plasmids and artificial chromosomes are used as vectors during recombinant DNA technology.

Artificial chromosomes are preferable to plasmids as vectors when larger fragments of foreign DNA are required to be inserted.

Restriction endonucleases cut open plasmids and cut specific genes out of chromosomes, leaving sticky ends.

Complementary sticky ends are produced when the same restriction endonuclease is used to cut open the plasmid and the gene from the chromosome.

Ligase seals the required gene into the plasmid.

Recombinant plasmids and artificial chromosomes contain restriction sites, regulatory sequences, an origin of replication and selectable markers.

Restriction sites contain target sequences of DNA where specific restriction endo-nucleases cut.

Regulatory sequences control gene expression and **origin of replication** allows self-replication of the plasmid/artificial chromosome.

Selectable markers such as antibiotic resistance genes protect the microorganism from a selective agent (antibiotic) that would normally kill it or prevent it growing.

Selectable marker genes present in the vector ensure that only microorganisms that have taken up the vector grow in the presence of the selective agent (antibiotic).

As a safety mechanism, **safety genes** are often introduced that prevent the survival of the microorganism in an external environment.

Plant or animal recombinant DNA expressed in bacteria may result in polypeptides being incorrectly folded and being non-functional and in these cases, recombinant yeast cells may be used to produce functional forms of the protein.

Key diagrams

Stages in the production of a genetically modified bacterial cell

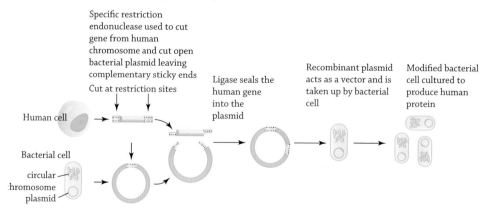

Features of a plasmid to be used in DNA technology

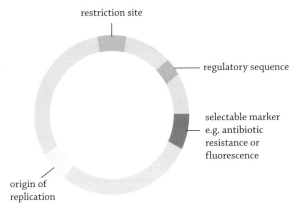

restriction site

regulatory sequence

selectable marker
e.g. antibiotic
resistance or
fluorescence

origin of
replication

> **!** Many candidates find difficulty in explaining the importance of using the same restriction endonuclease to remove the gene and open the plasmid – remember about complementary sticky ends.

> **!** Many candidates find difficulty in describing how to prevent survival of genetically modified microorganisms in the external environment – remember about including safety genes.

> **!** Many candidates find difficulty in describing how antibiotics can be used to select modified bacteria – remember that modified plasmids are often given selectable markers such as antibiotic resistance so that antibiotics can be used as selective agents to kill non-modified bacteria.

C-type questions

4. Bacterial plasmids are described as vectors.
   State what is meant by the term 'vector'.   1

5. Apart from plasmids, give **one** other example of a vector used in recombinant DNA technology.   1

6. Name **one** enzyme used in recombinant DNA technology and state its function.   2

7. Name the enzyme which can be used to seal the foreign DNA into the plasmid in recombinant DNA technology.   1

8. Give **two** features of a plasmid which allow it to be an effective vector.   2

9. Give an example of a suitable eukaryotic microorganism that can be used in genetic modification procedures instead of bacteria.   1

A-type questions

10. Bacteria are used in recombinant DNA technology.

    Explain why animal DNA which has been transferred to bacteria may produce proteins which are not functional. 1

11. Describe the role of a vector in recombinant DNA technology. 1

12. Describe **two** procedures that use restriction endonuclease enzymes in recombinant DNA technology. 2

13. Describe the role of genes introduced into a modified plasmid as a safety mechanism. 1

14. Explain why a plasmid must contain a restriction site to be an effective vector. 1

15. Explain the inclusion of a selectable marker gene in a recombinant plasmid. 1

16. Explain why a plasmid must contain an origin of replication to be an effective vector. 1

17. Describe the role of regulatory sequences in a modified plasmid. 1

18. Describe a situation in which the use of artificial chromosomes as vectors could be preferable to the use of plasmids. 1

19. State why yeast cells are often used in recombinant DNA technology in preference to bacterial cells. 2

Extended response questions

20. Describe the features required for a plasmid to function as an effective vector in recombinant DNA technology. 4

21. Give an account of recombinant DNA technology and how it is used in the production of proteins. 9

Model answers and commentary

Question		Model answer	Marks	Commentary with hints and tips
1		Mutagenesis; recombinant DNA technology	2	Mutagenesis is the process of inducing mutations.
2		UV light/(other forms of) radiation; mutagenic chemicals	2	Other forms of radiation which can induce mutation include X-rays.
3		Mutagenesis is the process of inducing mutations	1	Exposure to UV light, other forms of radiation or mutagenic chemicals results in random mutations, some of which might produce an improved strain of microorganism with desirable qualities.
4		A vector carries the DNA from the donor organism into the host cell	1	Recombinant plasmids or artificial chromosomes act as genetic vectors.

5		Artificial chromosome	1	These are preferable to plasmids when larger fragments of DNA are to be inserted.
6		Restriction endonuclease; enzyme that cuts genetic material from a chromosome/ is used to open a plasmid **OR** DNA ligase; seals gene/genetic material into plasmid	2	Treatment of vectors with the same restriction endonuclease forms complementary sticky ends that are then combined using DNA ligase.
7		DNA ligase	1	Combines the complementary sticky ends of the required gene with those of the plasmid.
8		Origin of replication **OR** restriction site **OR** selectable marker gene **OR** regulatory sequence **OR** genes preventing survival in external environment **(Any 2)**	2	This is just a list but reasons for these features would be needed to answer A-type questions well.
9		Yeast	1	Yeast is a special eukaryote – it has plasmids.
10		It can result in the production of polypeptides that are folded incorrectly	1	Bacterial cells may lack the ability to fold animal polypeptides correctly.
11		Carries foreign genetic information from the donor organism into the host cell	1	Recombinant plasmids and artificial chromosomes can act as vectors.
12		Restriction endonuclease enzyme cuts genetic material/ target sequences of DNA from a chromosome; Used to cut open a plasmid	2	Treatment with the same restriction endonuclease forms complementary sticky ends between the required gene and the plasmid, which are then combined using DNA ligase.
13		Genes are introduced that prevent the survival of a modified microorganism in an external environment	1	Vital in preventing release of bacteria with antibiotic resistance which they might pass on to wild populations of disease-causing bacteria by horizontal gene transfer. ⟹ **See 'What you need to know about the mechanism of evolution' in Chapter 7.**

14		These are locations on a DNA molecule (containing specific sequences of nucleotides) which are recognised by restriction enzymes	1	Restriction endonucleases cut target sequences/restriction sites of DNA, leaving sticky ends allowing the insertion of the required gene.	
15		A gene used to determine whether a DNA sequence/gene has been successfully inserted into the host organism's DNA	1	Marker genes include genes that give the bacteria resistance to an antibiotic or produce a green fluorescent protein which makes the modified cells glow green under UV light.	
16		This contains genes for the self-replication of the plasmid and ensures that the plasmid can be copied	1	Copied plasmid passed to daughter cells when the modified bacteria divide in culture allowing them to produce the desired protein.	
17		Control gene expression	1	Regulatory sequences/genes can increase transcription of a gene and increase the production of a protein.	
18		Preferable when larger fragments of foreign DNA need to be inserted	1	The chromosomes are larger than plasmids and can accommodate larger fragments of DNA.	
19		Yeast cells are eukaryotic; They have the additional enzymes/genes to fold the polypeptide correctly	2	Incorrectly folded or unfolded polypeptides are non-functional because they lack three-dimensional shape. ⟹ **See 'What you need to know about protein structure' in Chapter 3.**	
20	1	Restriction sites			4
	2	Selectable marker			
	3	Origin of replication/Genes for self-replication			
	4	Regulatory sequences/genes			
	5	Genes preventing survival in external environment **(Any 4)**			

21	1	Recombinant DNA technology involves the joining together of DNA molecules from two different species	9
	2	Genes/DNA sequences can be transferred to microorganisms to produce required proteins	
	3	Recombinant plasmids act as vectors	
	4	Artificial chromosomes act as vectors	
	5	The vector carries the foreign DNA from the donor organism into the host cell	
	6	Vectors must contain restriction sites	
	7	Vectors must contain selectable markers/genes for self-replication/ regulatory sequences	
	8	Restriction endonucleases cut target sequences of DNA, leaving sticky ends	
	9	Treatment of vectors with the same restriction endonuclease	
	10	Complementary sticky ends are then combined using DNA ligase to form recombinant DNA	
	11	Genes to prevent the survival of the microorganism in an external environment can be introduced as a safety mechanism **(Any 9)**	

Food supply, plant growth and productivity

What you need to know about food supply and sustainable food production

Food security is the ability of human populations to access food of sufficient quality and quantity.

Increase in human population and concern for food security lead to a demand for increased food production.

Food production must be sustainable and not degrade the natural resources on which agriculture depends.

Agricultural production depends on factors that control photosynthesis and plant growth.

The area to grow crops is limited.

Increased food production will depend on factors that control plant growth, including the breeding of higher-yielding **cultivars**, the use of fertiliser and protecting crops from **pests**, diseases and competition.

All food production is dependent ultimately upon photosynthesis.

Plant crop examples include cereals, potato, roots and legumes.

Breeders seek to develop crops with higher nutritional values, resistance to pests and diseases, physical characteristics suited to rearing and harvesting as well as those that can thrive in particular environmental conditions.

Livestock produce less food per unit area than crop plants due to loss of energy between **trophic levels**.

Livestock production is often possible in habitats unsuitable for growing crops.

➡ **There are more details about these broad ideas in Chapters 17 and 18.**

Key diagram

Plants produce more food per unit area for humans **(a)** than livestock **(b)** because of the loss of energy between trophic levels.

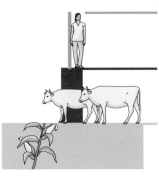

Secondary consumer trophic level
Consumers with only 1 unit of **energy** because energy has been lost between the trophic levels

Primary consumer tropic level
Consumers with only 10 units of **energy** because energy has been lost between the trophic levels

Producer trophic level
Crop plants with 100 units of energy containing energy trapped during photosynthesis

C-type questions

1. Give the reason for the need to increase global food production.   1
2. Give **three** examples of methods of increasing yield of food from a crop plant such as wheat.   3
3. State what is meant by the following terms:
    (a) 'photosynthesis'   1
    (b) 'trophic level'   1

A-type questions

4. Explain what is meant by the term 'food security'.   3
5. Explain why there has been a major increase in concern about food security in recent years.   2
6. Describe how **two** methods of improving food production from crops lead to increases in yield.   2
7. Plants can be grown for human food or used as food for livestock.

    In terms of food security, explain **one** benefit of using plants rather than livestock for human food.   2

➡ **Model answers and commentary can be found on page 125.**

What you need to know about light and photosynthesis

Light energy is absorbed by photosynthetic pigments to generate ATP and for **photolysis**.

Light energy which is not absorbed is transmitted or reflected.

Absorption spectra of chlorophyll a and b and carotenoids should be compared to **action spectra** for photosynthesis.

Carotenoids extend the range of wavelengths absorbed and pass the energy to chlorophyll for photosynthesis.

Each pigment absorbs a different range of wavelengths of light.

Absorbed light energy excites electrons in the pigment molecule.

Key diagram

Action spectrum of a green plant and absorption spectrum of chlorophyll a – notice that although chlorophyll does not absorb well in the blue-to-green and the yellow-to-orange areas of the spectrum, photosynthesis is still occurring due to the absorption of light by other pigments, including carotenoids.

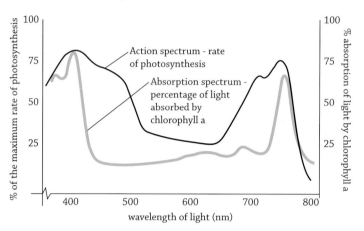

Technique

Using a **spectroscope** is a technique you need to be familiar with for your exam and one that could be used in your assignment.

A spectroscope is an instrument which is used to look at the spectrum of light coming from a source. For example, looking at light which has passed through a chlorophyll solution – the visible spectrum in this case would have only green light in it because the other colours would be absorbed by the chlorophyll solution.

Technique

Carrying out **chromatography** is a technique you need to be familiar with for your exam and one that could be used in your assignment.

Chromatography is a method for the separation of different coloured pigments in a plant extract. The separation can occur using paper or in an adsorbant layer of silica gel on glass or foil.

C-type questions

8. Apart from being absorbed, give **one** other fate of light which strikes the leaves of plants.   1

9. Apart from chlorophyll a, name a pigment or group of pigments which can extend the range of wavelengths of light absorbed.   1

A-type questions

10. Explain the difference between the absorption spectrum of a pigment and the action spectrum of a green plant.   2

11. Describe the part played by carotenoid pigments in photosynthesis.   2

Many candidates have difficulty comparing absorption and action spectra – remember that action spectra show rates of photosynthesis.

➡ **Model answers and commentary can be found on page 126.**

What you need to know about the stages of photosynthesis

Transfer of electrons through the electron transport chain releases energy to generate ATP by **ATP synthase**.

Energy is also used for photolysis, in which water is split into oxygen, which is evolved, and hydrogen, which is transferred to the **coenzyme NADP**.

In the carbon fixation stage (**Calvin cycle**), the enzyme **RuBisCO** fixes carbon dioxide by attaching it to **ribulose bisphosphate (RuBP)**.

The **3-phosphoglycerate (3PG)** produced is phosphorylated by ATP and combined with hydrogen from NADPH to form **glyceraldehyde-3-phosphate (G3P)**.

G3P is used to regenerate RuBP and for the synthesis of glucose.

Glucose may be used as a respiratory substrate, synthesised into starch or cellulose or passed to other biosynthetic pathways.

Biosynthetic pathways can lead to the formation of a variety of metabolites such as DNA, protein and fat.

Key diagrams

Light-dependent stage of photosynthesis in a membrane within a chloroplast

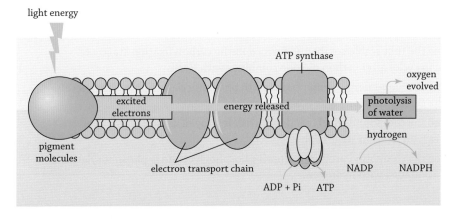

The carbon fixation stage (Calvin cycle) in which carbon dioxide from the air is fixed into glucose

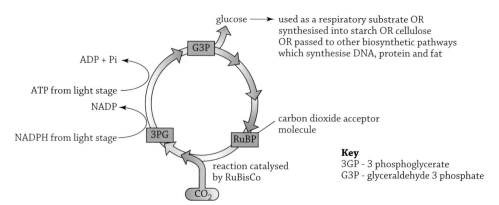

Many candidates have trouble relating pigments absorbing light energy to transfer of electrons down electron transport chains – remember that light energy excites electrons in pigment molecules.

Many candidates find it tricky to describe events in the carbon fixation stage (Calvin cycle) – remember that using the abbreviations is acceptable and might make it easier to remember.

C-type questions

12. Name **two** substances formed in the light-dependent stage which are required for the carbon fixation stage (Calvin cycle). 2

13. Name the coenzyme that transports hydrogen to the carbon fixation stage (Calvin cycle). 1

14. Name the compound which accepts hydrogen from the photolysis of water. 1

15. Name the molecule which acts as the carbon dioxide acceptor in the carbon fixation stage (Calvin cycle). 1

16. Name the enzyme that attaches carbon dioxide to RuBP. 1

17. Name the first stable compound formed in the carbon fixation stage (Calvin cycle) after carbon dioxide combines with RuBP. 1

18. Name the high-energy compound used to phosphorylate 3-phosphoglycerate in the carbon fixation stage (Calvin cycle) 1

19. Name the compound which combines with hydrogen during the carbon fixation stage (Calvin cycle) 1

20. Name the compound produced after 3-phosphoglycerate is phosphorylated and combined with hydrogen. 1

A-type questions

21. Light energy absorbed by photosynthetic pigments excites electrons in the pigment molecules of the chloroplast.

    Describe the role of the excited electrons in photosynthesis.   2

22. During photosynthesis, G3P is produced when an intermediate substance is combined with hydrogen and phosphorylated by ATP.

    Name the stage of photosynthesis which provides the hydrogen and ATP for this reaction and the intermediate substance involved.   2

23. Explain how the conversion of 3-phosphoglycerate to G3P in the carbon fixation stage (Calvin cycle) is dependent on chemical reactions in the light-dependent stage of photosynthesis.   2

24. Describe the fate of G3P produced during the carbon fixation stage (Calvin cycle).   2

25. Describe **three** fates of the glucose produced by photosynthesis.   3

26. Describe the changes in concentration of RuBP and G3P that would be expected to occur if an illuminated green plant cell's source of carbon dioxide were removed.   1

Extended response questions

27. Give an account of food security and sustainability.   5

28. Describe the role of photosynthetic pigments in the process of photosynthesis.   6

29. Give an account of the events of the carbon fixation stage (Calvin cycle) in photosynthesis.   4

30. Write notes on human food supply under the following headings:
    (a) food security and population   3
    (b) factors affecting food production   6

31. Give an account of photosynthesis under the following headings:
    (a) photolysis   4
    (b) carbon fixation stage (Calvin cycle)   5

Model answers and commentary

Question		Model answer	Marks	Commentary with hints and tips
1		Population increase	1	Human population growth has increased the demand for food and the concern for food security.
2		Plant a greater area of crop; use fertilisers; use pesticide; use biological control methods; grow improved/genetically modified (GM) strains/cultivars of crop plants (other answers possible) **(Any 3)**	3	The area to grow crops is limited. Increased food production depends on factors that control plant growth – breeding of higher-yielding cultivars, use of fertiliser, protecting crops from pests, diseases and competition.

3	(a)	Production of food/carbohydrate by a green plant using the energy of light	1	Photosynthesis is a process in which green plants trap light energy and use it in the production of food.
	(b)	An organism's feeding level/ position in a food chain	1	Humans can occupy different trophic levels. Due to the loss of energy between trophic levels, it is more energy efficient for humans to act as primary consumers and eat crops than to eat livestock as secondary consumers.
4		The ability to access sufficient quality; and quantity of food; over a sustained/extended period of time	3	Remember: Food security = **QQAS** **Q**uantity, **Q**uality, **A**ccessibility and **S**ustainability.
5		Increase in global/world/human population; requires increased food production	2	Land area is a limiting factor.
6		Plant a greater area of crop; use fertilisers; use pesticide; use biological control methods; grow improved/GM strains/cultivars of crop plants; ensuring good supply of water; good soil management to avoid erosion **(Any 2)**	2	Agricultural production depends on factors that control photosynthesis and plant growth.
7		Plants/crops are producers and livestock are consumers **OR** crops and livestock are at different trophic levels; energy is lost between each trophic level of a food chain so greater number of people could be fed using crops	2	Energy is lost between each trophic level. It is more energy efficient for humans to act as primary consumers and eat crops than to eat livestock as secondary consumers.
8		Reflected **OR** transmitted	1	Remember 'ART': **A**bsorbed, **R**eflected and **T**ransmitted. Increased difficulty by giving you one example.
9		Chlorophyll b/carotenoids	1	Each pigment absorbs a different range of wavelengths of light.
10		An absorption spectrum shows the extent to which each wavelength of light is absorbed by a pigment; An action spectrum shows the rate of photosynthesis by a whole plant at each wavelength	2	These graphs are often shown superimposed, as in the Key diagram, to show the role of the carotenoid pigments.

11		Carotenoid pigments extend the range of wavelengths of light absorbed for photosynthesis; and pass the energy onto chlorophyll	2	The carotenoids absorb violet and blue–green light.
12		NADPH; ATP	2	ATP is used to phosphorylate 3PG. NADPH supplies the hydrogen which is combined with 3PG to produce G3P.
13		NADP	1	You need to know that NADP is a coenzyme.
14		NADP	1	Rewording of previous question.
15		RuBP	1	RuBP combines with CO_2 to produce 3PG.
16		RuBisCo	1	RuBisCo fixes CO_2 from the air by attaching it to RuBP to form 3PG.
17		3PG/3-phosphoglycerate	1	3PG is then converted to G3P by the addition of hydrogen.
18		ATP	1	Remember, in respiration, ATP is used in the phosphorylation of glucose and intermediates during glycolysis and here in the phosphorylation of 3PG in photosynthesis.
19		3PG	1	3PG is then converted to G3P by the addition of hydrogen.
20		G3P/glyceraldehyde-3-phosphate	1	G3P is used to regenerate RuBP and for the synthesis of glucose.
21		Release energy; to generate ATP by ATP synthase	2	Transfer of the excited electrons through the electron transport chain releases energy to generate ATP by ATP synthase.
22		Light-dependent stage; 3PG/3-phosphoglycerate	2	These questions need you to have a good understanding of the stages of photosynthesis. Practise writing out these stages until you are perfect.
23		Light-dependent stage produces the ATP; and NADPH required to reduce 3PG/convert 3PG into G3P	2	These questions need you to have a good understanding of the stages of photosynthesis. Practise writing out these stages until you are perfect.

24		The G3P is converted into glucose; or can be used to regenerate RuBP	2	Students often forget that G3P is also used to regenerate RuBP to continue the cycle.
25		The glucose produced can be used in respiration/as a respiratory substrate; synthesised/converted to cellulose; synthesised/converted into starch; used in the synthesis of a variety of metabolites/DNA/proteins/fat **(Any 3)**	3	'Fate' means what can happen to the glucose or how it is used.
26		RuBP would increase **AND** G3P would decrease	1	Check the Calvin cycle diagram and see the effect of the reduction of the CO_2.
27	1	Increasing population increases the demand for food production		5
	2	Food security is ability of human population to access sufficient quantity/amount of food		
	3	Sufficient quality of food		
	4	Ability to distribute/spread food through the population		
	5	Production must be guaranteed over a long period of time		
	6	Food production should not degrade natural resources on which agriculture depends **(Any 5)**		
28	1	Photosynthetic pigments are located in the chloroplasts of plants		6
	2	Main pigments are chlorophyll a and chlorophyll b		
	3	Pigments absorb light energy		
	4	Light energy absorbed by chlorophyll a excites electrons into a high-energy state		
	5	Electrons pass through an electron transport chain		
	6	ATP synthase produces/synthesises ATP		
	7	Some light energy absorbed by pigments is used to split a molecule of water into hydrogen and oxygen		
	8	This process is called photolysis		
	9	ATP and NADPH are produced for the Calvin cycle/carbon fixation stage **(Any 6)**		
29	1	Carbon dioxide joined to RuBP by RuBisCO		4
	2	To produce 3-phosphoglycerate/3PG		
	3	ATP used to phosphorylate 3-phosphoglycerate/3PG to form G3P/glyceraldehyde-3-phosphate		
	4	Hydrogen from NADPH used to form G3P		
	5	G3P forms glucose		
	6	Some G3P regenerates RuBP **(Any 4)**		

30	(a)	1	Increasing population increases the demand for food production	3
		2	Food security is ability of human population to access sufficient quantity/amount of food	
		3	Sufficient quality of food	
		4	Ability to distribute/spread food through the population	
		5	Production must be guaranteed over a long period of time	
		6	Food production should not degrade natural resources on which agriculture depends **(Any 3)**	
	(b)	1	Food production depends on photosynthesis	6
		2	Food production can be increased by planting increased area of crop	
		3	Factors which limit photosynthesis/control plant growth include light/temperature/CO_2 availability	
		4	High-yielding cultivars/GM crops	
		5	Protection of crops from disease/pests/competition	
		6	Increased irrigation/fertilisers	
		7	Livestock produce less food per unit area than crops	
		8	Due to energy loss at each trophic level **(Any 6)**	
31	(a)	1	Light energy excites electrons in pigment	4
		2	Splitting of water molecules in light-dependent stage	
		3	Using energy from electron	
		4	Oxygen produced which is released	
		5	Hydrogen is accepted by NADP **(Any 4)**	
	(b)	1	Carbon dioxide joined to RuBP	5
		2	RuBisCo catalyses this reaction	
		3	To produce 3-phosphoglycerate/3PG	
		4	ATP used to phosphorylate 3-phosphoglycerate/3PG to form G3P/ glyceraldehyde-3-phosphate	
		5	Hydrogen from NADPH used to form G3P	
		6	G3P forms glucose	
		7	Some G3P regenerates RuBP **(Any 5)**	

CHAPTER 17   Plant and animal breeding

What you need to know about the improved characteristics sought by breeding programmes

Plant and animal breeding is carried out to improve characteristics to help support sustainable food production.

Breeders develop crops and animals with higher food yields, higher nutritional values, pest and disease resistance and the ability to thrive in particular environmental conditions.

Key diagram

Summary of some characteristics desired by animal and plant breeders

Improved characteristics	Example
Higher food yield	High milk yielding cattle
Higher nutritional value	Golden rice with enhanced vitamin A content
Pest and disease resistance	Blight resistance in potatoes
Environmental tolerances such as drought, frost, salinity, flooding	Frost resistance in strawberries and drought resistance in wheat

C-type question

1. Give **two** characteristics that might be selected by breeders seeking to improve a crop plant species.   2

A-type question

2. Explain the role of breeders in supporting sustainable food production.   2

➡ **Model answers and commentary can be found on page 135.**

What you need to know about field trials

Plant **field trials** are carried out in a range of environments to compare the performance of different cultivars or treatments and to evaluate **genetically modified (GM) crops**.

In designing field trials account has to be taken of the selection of treatments, the number of **replicates** and the **randomisation** of treatments.

The **selection of treatments** must ensure valid comparisons between cultivars, the number of replicates involved must take account of the variability within the sample and the randomisation of treatments is needed to eliminate bias when measuring treatment effects.

Key diagram

A field trial involving different cultivars of rice plants

C-type questions

3. Give the term applied to a non-laboratory test carried out to evaluate the performance of different cultivars in various environmental conditions.   1

4. GM crops are evaluated in field trials. Certain experimental procedures are required when setting up field trials to compare GM and non-GM crops.

    Give **one** such procedure and explain how it allows valid conclusions to be drawn.   2

A-type questions

5. Explain the purpose of carrying out plant field trials.   1

6. Explain the need for replication and randomisation of treatments in a plant breeding field trial.   2

➡ **Model answers and commentary can be found on page 135.**

What you need to know about inbreeding

In **inbreeding**, selected related plants or animals are bred for several generations until the population **breeds true** to the desired type due to the elimination of heterozygotes.

A result of inbreeding can be an increase in the frequency of individuals which are **homozygous** for **recessive** deleterious alleles.

Homozygous individuals will do less well at surviving to reproduce resulting in **inbreeding depression**.

Key diagram

Summary of inbreeding

Method	Advantages	Disadvantages to breeder
Breed related individuals for several generations	Produce true-breeding varieties and eliminate heterozygotes giving homozygous offspring of the desired type	Increase in individuals homozygous for recessive deleterious alleles and showing inbreeding depression though poor reproductive rates and survival frequency

➡ **There is more about inbreeding depression in Chapter 23.**

C-type questions

7. Give the term used to describe the process of crossing related individuals of similar genotypes until the population breeds true due to the elimination of heterozygotes.   1

8. Give the term used to describe the deleterious effects of the breeding together of genetically similar individuals.   1

A-type question

9. Explain the meaning of the term 'inbreeding depression'.   1

➡ **Model answers and commentary can be found on page 136.**

What you need to know about crossbreeding and F_1 hybrids

In animals, individuals from different breeds may produce a new **crossbreed** population with improved characteristics.

The two parent breeds can be maintained to produce more crossbred animals showing the improved characteristic.

New alleles can be introduced to plant and animal lines by crossing a cultivar or breed with an individual with a different, desired genotype.

In plants, F_1 hybrids, produced by the crossing of two different inbred lines, create a relatively uniform **heterozygous** crop.

F_1 hybrids often have increased vigour and yield.

Plants with increased vigour may have increased disease resistance or increased growth rate.

When breeding animals and plants, F_1 hybrids are not usually bred together as the F_2 produced shows too much variation.

Key diagram

Summary of crossbreeding

---

1   Parent breed 1   ×   Parent breed 2
                        ↓

F_1 hybrid – improved breed which combines desired characteristics of parents and may have **hybrid vigour**, which involves increased yield, disease resistance and increased growth rate

2 F_1 hybrid   × F_1 hybrid
                      ↓

F_2 population – would show too much variation so this cross not usually undertaken

---

C-type questions

10. Give the reason why individuals from different breeds are sometimes crossed.   1

11. In plant breeding, F_1 hybrids of different varieties of a species are often produced because they combine desired features of their parent varieties.

    Explain why F_2 plants produced from the hybrid are considered of little use for further production.   1

> **!** Candidates often have difficulty explaining why F_1 hybrids are not used to produce an F_2 generation – remember that any F_2 produced show too much variation and a lower percentage show the hybrid vigour desired.

A-type question

12. Describe **one** method of maintaining a new crossbreed population that shows improved characteristics.   1

➡ **Model answers and commentary can be found on page 136.**

What you need to know about genetic technology

As a result of **genome sequencing**, organisms with desirable genes can be identified and then used in breeding programmes.

Single genes for desirable characteristics can be inserted into the genomes of crop plants, creating genetically modified plants with improved characteristics.

Breeding programmes can involve crop plants that have been genetically modified using **recombinant DNA technology**.

Recombinant DNA technology in plant breeding includes insertion of the **Bt toxin gene** into plants for pest resistance and the **glyphosate resistance gene** inserted for herbicide tolerance.

Key diagram

Example of recombinant DNA technology – a bacterial gene inserted into a corn plant

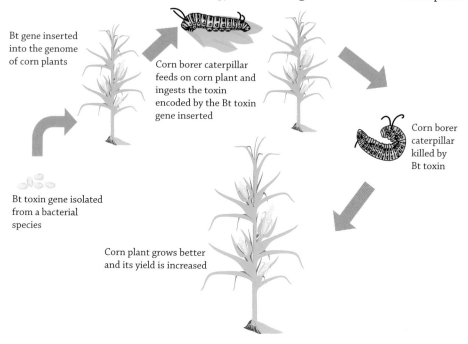

Bt gene inserted
into the genome
of corn plants

Corn borer caterpillar
feeds on corn plant and
ingests the toxin
encoded by the Bt toxin
gene inserted

Corn borer
caterpillar
killed by
Bt toxin

Bt toxin gene isolated
from a bacterial
species

Corn plant grows better
and its yield is increased

➡ **There is more about genetic modification in Chapter 15.**

C-type questions

13. Name the procedure carried out to identify organisms with genes for desirable characteristics so they can be used in breeding programmes.   1

14. Give **one** example of an application of recombinant DNA technology in plant breeding.   1

A-type questions

15. Describe the role of genomic sequencing in animal breeding programmes.   1

16. Describe the role of genomic sequencing in breeding programme to improve crop plants.   2

17. Explain the advantage of producing a plant cultivar which has had a Bt toxin gene inserted into its genome.   2

18. Explain the advantage of producing a plant cultivar which has had the glyphosate resistance gene inserted into its genome.   2

Extended response questions

19. Give an account of inbreeding **and** crossbreeding and outline the likely effects of carrying out these breeding approaches.   8

20. Give an account of plant field trials in the evaluation of GM crop cultivars.   7

Model answers and commentary

Question		Model answer	Marks	Commentary with hints and tips
1		Higher yield; higher nutritional value; resistance to pests/diseases/herbicides; characteristics suited to growing/harvesting; characteristics suitable for survival in particular habitats/environmental conditions **(Any 2)**	2	Any useful feature that would contribute to increasing food security or reduce the dependency on chemicals.
2		Breeders develop crops/animals; With higher food yields/higher nutritional values/pest resistance/disease resistance/ability to thrive in particular environmental conditions	2	Involves manipulating heredity to improve new crops and animal stock.
3		Field trial	1	Carried out to test new pesticides, seed varieties/cultivars and fertilisers in a variety of locations and conditions.
4		Replication; takes account of the variability within the sample and increases the reliability of the results **OR** Randomisation; eliminates bias/effects of factors other than the treatment when measuring the treatment's effects	2	Remember '**RR**' for field trials: **R**eplication and **R**andomisation.
5		To compare the performance of plots of different varieties of crop plant/cultivars **OR** To compare how plots of one cultivar perform with a range of treatments/different fertiliser application levels	1	Can be designed to test promising new plants in a situation similar to the actual growing conditions by farmers, such as unpredictable light and temperature, the presence of unknown microbes and animals in the soil, and competing weeds.

6		Replication takes account of the variability within the sample and increases the reliability of the results; Randomisation eliminates bias/effects of factors other than the treatment when measuring the treatment's effects	2	Tricky definitions that just need to be learned. Use the flash card glossary.
7		Inbreeding	1	Breeding related individuals to increase the frequency of individuals that are homozygous for a desired characteristic/allele.
8		Inbreeding depression	1	When inbreeding has resulted in harmful/unwanted recessive alleles becoming homozygous and resulting in offspring with reduced biological fitness.
9		An increase in the effects of recessive deleterious alleles due to inbreeding	1	Deleterious = harmful or unfavourable.
10		Crossbreeding individuals from different breeds can produce a new F_1 crossbreed population/individuals with improved characteristics/yield/hybrid vigour/increased disease resistance/increased growth rate	1	Crossbreeds are not usually bred through to the F_2 generation because these would show too much variation – more crossbreeds are produced by further breeding of the parental types.
11		As the F_2 produced shows too much variation	1	The F_2 produced have a lower percentage showing the improved characteristic/hybrid vigour.
12		The two parent breeds can be maintained to produce more crossbred animals showing the improved characteristic	1	Parent breeds are usually maintained by specialised breeders.
13		Genomic sequencing	1	Carried out to determine the order of DNA nucleotides/bases in a genome.
14		Insertion of Bt toxin gene into plants for pest resistance **OR** glyphosate resistance gene inserted for herbicide tolerance	1	Less pesticide is needed if a crop produces its own pesticide chemical/toxin. Herbicide-resistant crops can be treated with the weed killer to remove competing weeds without being damaged themselves.

15		Individuals with desirable genes can be identified and then used in breeding programmes	1	A faster and more precise technique/method that allows genetic material from one species to be identified and then inserted into the genome of another.	
16		Single genes identified for desirable characteristics can be inserted into the genomes of crop plants; creating genetically modified plants with improved characteristics	2	The genetic material of the transformed organism is undisturbed apart from the insertion of a gene.	
17		Produces its own pesticide/toxin; Increases yield without need for pesticide application	2	**There is more about problems with pesticides in Chapter 18.**	
18		Produces crop/plant with herbicide tolerance; Allows use/treatment with herbicide to remove/kill competing weeds but not the cultivar	2	**There is more about herbicides in Chapter 18.**	
19		Inbreeding:			8
	1	Select related individuals/plants/animals to breed			
	2	For several generations			
	3	Until population breeds true to desired type			
	4	Less/no heterozygotes			
	5	Can result in inbreeding depression			
	6	Increase in homozygotes/individuals homozygous for recessive deleterious/harmful alleles			
	7	Poor reproductive rates/survival frequencies **(Any 5)**			
	8	Crossbreeding: Cross two selected breeds with different desired characteristics			
	9	F_1 hybrids combine these characteristics			
	10	F_1 show hybrid vigour/increased yield/increased growth rate/increased disease resistance			
	11	(Not usually crossed together/F_2 not produced) because they show too much variation **(Any 3)**			

20	1	Carried out in a range of environments	7
	2	Carried out to compare the performance of different cultivars	
	3	Different treatments	
	4	Account has to be taken of the selection of treatments	
	5	The number of replicates	
	6	The randomisation of treatments	
	7	The selection of treatments to ensure valid comparisons	
	8	The number of replicates to take account of the variability within the sample	
	9	The randomisation of treatments to eliminate bias when measuring treatment effects **(Any 7)**	

18   Crop protection

What you need to know about crop plant pests

Weeds compete with crop plants, while other **pests** and diseases damage crop plants, all of which reduce productivity.

Properties of **annual weeds** include rapid growth, short life-cycle, high seed output and long-term seed viability.

Properties of **perennial weeds** with competitive adaptations include storage organs and vegetative reproduction.

Most of the pests of crop plants are invertebrate animals such as insects, nematode worms and molluscs.

Plant diseases can be caused by fungi, bacteria or viruses, which are often carried by invertebrates.

Key diagram

Summary of the pests of crop plants

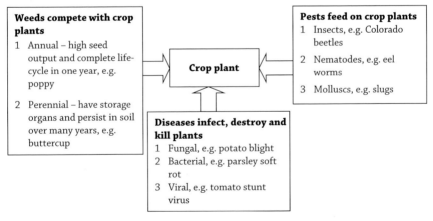

Weeds compete with crop plants
1 Annual – high seed output and complete life-cycle in one year, e.g. poppy
2 Perennial – have storage organs and persist in soil over many years, e.g. buttercup

Crop plant

Pests feed on crop plants
1 Insects, e.g. Colorado beetles
2 Nematodes, e.g. eel worms
3 Molluscs, e.g. slugs

Diseases infect, destroy and kill plants
1 Fungal, e.g. potato blight
2 Bacterial, e.g. parsley soft rot
3 Viral, e.g. tomato stunt virus

C-type questions

1. Name **two** types of organism that cause plant diseases.   2
2. Name **two** invertebrate animal groups which can be pests of crop plants.   2
3. Give **one** property of an annual weed.   1
4. Give **one** property of a perennial weed.   1

A-type questions

5. Describe the adaptations of annual plants that allow them to be successful weeds of crop plants.   2
6. Describe the adaptations of perennial plants that allow them to be successful weeds of crop plants.   2
7. Describe **two** differences between annual and perennial weeds.   2

➡ **Model answers and commentary can be found on page 143.**

What you need to know about cultural and chemical methods of pest control

Weeds, other pests and diseases can be controlled by **cultural methods** that include ploughing, weeding and crop rotation.

Chemical methods involve **pesticides** which include herbicides to kill weeds, **fungicides** to control fungal diseases, insecticides to kill insect pests, molluscicides to kill mollusc pests and nematicides to kill nematode pests.

Selective herbicides have a greater effect on certain plant species (broad-leaved weeds).

Systemic herbicides spread through the vascular system of plants and prevent regrowth.

Systemic insecticides, molluscicides and nematicides spread through the vascular system of plants and kill pests feeding on plants.

Problems with pesticides include toxicity to non-target species, **persistence** in the environment and **bioaccumulation** or **biomagnification** in food chains, producing resistant populations of pests.

Applications of fungicide based on **disease forecasts** are more effective than treating diseased crops.

Bioaccumulation is a build-up of a chemical in an organism.

Biomagnification is an increase in the concentration of a chemical moving between trophic levels.

Key diagrams

Summary of the effects of different cultural methods of crop protection

Cultural method	Effect
Ploughing	Perennial weeds damaged or buried so cannot grow and compete with crop for light, water, nutrients and space
Weeding	Manual removal of annual weeds so that they do not compete with the crop for light, water and nutrients
Crop rotation	Growing different crops in rotation in a field to prevent crop-specific pest species from building up

Summary of the effects of chemical methods on pests

Chemical method	Effect
Selective herbicide	Targets broad-leaved plant weeds
Systemic herbicide	Spreads through vascular system of plants and prevents regrowth
Systemic insecticide, molluscicide and nematicide	Spreads through vascular system of plants and kills insects, molluscs and nematodes feeding on crop plants
Fungicide	Controls fungal diseases

Summary of the problems with pesticides

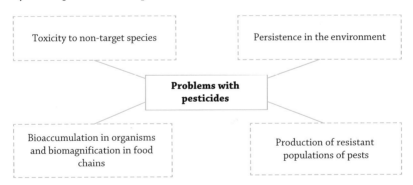

C-type questions

8. Give **one** example of a cultural method of controlling weeds, pests or diseases.   1
9. Give the name for the pesticides that can target broad-leaved species of weeds.   1
10. Give the name for the pesticides that are effective against weeds with underground storage organs.   1
11. Give **two** problems associated with pesticides.   2

A-type questions

12. Describe how cultivation methods can be used to control the growth of weeds.   2
13. Describe the action of selective herbicides.   1

14. Describe the action of systemic herbicides.   2
15. Describe **two** environmental problems associated with applying insecticide to crops.   2
16. Explain the difference between a selective herbicide and a systemic herbicide.   1
17. Describe the use of disease forecasts in crop protection.   1
18. Explain why systemic herbicides are effective against weeds with underground storage organs.   2
19. Explain the meaning of the term 'persistent pesticide'.   2
20. Explain the difference between bioaccumulation and biomagnification.   2
21. Explain how application of a pesticide can produce a resistant population of a pest.   4

➡ **There is more about selection pressure in Chapter 7.**

➡ **Model answers and commentary can be found on page 144.**

What you need to know about biological control and integrated pest management

Weeds and other pests and diseases can also be controlled by using **biological control** and **integrated pest management (IPM)**.

In biological control the control agent is a natural predator, parasite or pathogen of the pest.

Integrated pest management is a combination of chemical, biological and cultural control.

A risk associated with biological control is that the control organism may become an invasive species, parasitise, prey on or be a pathogen of other species.

Key diagram

Examples of biological control methods

Biological control agent	Example
Predator	Introduce ladybirds to a greenhouse to predate greenfly so preventing damage to a crop
Parasite	Introduce *Encarsia* wasps, which lay eggs on whiteflies and so prevent them damaging a crop
Pathogen	Introduce the pathogenic bacteria *Bacillus thuringiensis*, which infects caterpillars so preventing them eating the crop

➡ **There is more about invasive species in Chapter 23.**

C-type questions

22. Identify a possible risk to a food web which could be associated with the use of a biological control method.   1
23. Name **two** different control methods used in IPM.   2

A-type questions

24. Describe the effects of **one** biological control agent in the control of a pest.   1

25. Explain the meaning of the term 'integrated pest management' (IPM).   1

26. Explain how a biological control agent might become invasive.   2

Extended response questions

27. Give an account of biological control of pests and outline a risk associated with this method of crop plant protection.   3

28. Give an account of chemical methods used to protect plants and of the environmental damage which may result from their use.   9

29. Give an account of the effects of weeds, pests and diseases on crop plant yield.   8

30. Give an account of methods used to protect crop plants.   6

Model answers and commentary

Question		Model answer	Marks	Commentary with hints and tips
1		Fungi; bacteria; viruses **(Any 2)**	2	'The usual suspects' often carried by invertebrates.
2		Insects; nematodes; molluscs **(Any 2)**	2	These can damage crops directly or infect crops with disease.
3		Rapid growth; short life-cycle; high seed output; long-term seed viability **(Any 1)**	1	Annual = complete life-cycle within a year. Think sexual reproduction, flowers and seeds for these weeds.
4		Storage organs **OR** vegetative reproduction	1	Perennial = lives for two years or more. To do this they need underground food stores to be used for next year's growth. Vegetative reproduction = any form of asexual reproduction.
5		Rapid growth; short life-cycle **OR** complete life-cycle within a year; high seed output; long-term seed viability **(Any 2)**	2	Any question referring to features, competitive adaptations, properties and success in these plants simply requires you to be able to list these properties.
6		Storage organs; vegetative reproduction; early growth; larger size **(Any 2)**	2	Any question referring to features, competitive adaptations, properties and success in these plants simply requires you to be able to list these properties.

7		Annual plants have a short life-cycle/complete life-cycle within a year; Perennial plants are plants that live for two years or more **OR** Annual plants have a high seed output/long-term seed viability; Perennial plants have storage organs/vegetative reproduction	2	If you can list their features, you can do this comparison.
8		Ploughing; weeding; crop rotation **(Any 1)**	1	Traditional methods often involving farming practice which do not involve chemicals.
9		Selective herbicides	1	Selective = targets broad-leaved species of weeds, leaving the desired crop relatively unharmed.
10		Systemic herbicides	1	Systemic herbicide spreads through the vascular system of the plant (phloem) and prevents regrowth.
11		Toxicity to non-target species; persistence in the environment; bioaccumulation/biomagnification in food chains; producing resistant populations of pests **(Any 2)**	2	Bioaccumulation is a build-up of a chemical in an organism. Biomagnification is an increase in the concentration of a chemical moving between trophic levels.
12		Ploughing; weeding; crop rotation **(Any 2)**	2	Traditional methods often involving farming practice which do not involve chemicals.
13		Selective herbicides have a greater effect on broad-leaved weeds	1	Selective = targets broad-leaved species of weeds, leaving the desired crop relatively unharmed.
14		Systemic herbicide spreads through vascular system of plant; and prevents regrowth	2	Systemic herbicides are transported through the phloem and can reach and kill underground organs and roots.

15		Toxicity to non-target species; persistence in the environment/bioaccumulation/ biomagnification in food chains; production of resistant populations of pests **(Any 2)**	2	Bioaccumulation is a build-up of a chemical in an organism. Biomagnification is an increase in the concentration of a chemical moving between trophic levels.
16		Selective herbicides have a greater effect on broad-leaved weeds **AND** Systemic herbicides spread through the vascular system of the plant	1	Selective = targets broad-leaved species of weeds. **System**ic herbicides are transported through the plant's transport **system**.
17		Applications of fungicide based on disease forecasts are more effective than treating diseased crops	1	The disease forecast takes into account factors such as the type of crop, the pathogen and the weather conditions likely to cause an infection.
18		Systemic herbicide spreads through the vascular system of the plant and reaches the underground storage organs; and prevents regrowth/ regeneration	2	Systemic herbicides are transported through the phloem and can reach and kill underground organs and roots.
19		Chemicals that will not breakdown/biodegrade in the environment; bioaccumulation/ biomagnification	2	Bioaccumulation is a build-up of a chemical in an organism. Biomagnification is an increase in the concentration of a chemical moving between trophic levels.
20		Bioaccumulation is the build-up of toxic chemicals in the body of an individual ingesting the chemical in its food; Biomagnification is the overall effect of bioaccumulation along a food chain	2	Try linking biomagni**F**ication with **F**ood chain.
21		Pesticide acts as a selection pressure; many individuals die; any individual with a mutation which gives them resistance survive; they reproduce and pass their resistance on to offspring	4	➡ **Quite tricky – read more about selection pressures in Chapter 7.**

22		The control organism may become an invasive species/parasitise/prey on/be a pathogen of/outcompete/hybridise with native species	1	➡ **Also covered in 'What you need to know about introduced, naturalised and invasive species' in Chapter 23.**
23		Chemical; biological; cultural **(Any 2)**	2	Integrated = a combination of.
24		In biological control the control agent is a natural predator **OR** parasite **OR** pathogen of the pest	1	Remember '**PPP**': **P**redator, **P**athogen or **P**est.
25		Integrated pest management is a combination of chemical, biological and cultural control	1	Integrated = a combination.
26		Introduced species may escape into a natural community; might outcompete/predate/hybridise with native species	2	➡ **Read more about invasive species in Chapter 23.**

27	1	Use of natural predator/parasite/disease to control pest numbers	3
	2	Better used in enclosed situation/greenhouse	
	3	Introduced organisms may be free of predators/parasites/disease themselves	
	4	Could become invasive/a threat to indigenous/native species **(Any 3)**	
28	1	Any example from herbicide/insecticide/fungicide	9
	2	Another example	
	3	Herbicides can be systemic and pass through the plant transport system/phloem	
	4	Herbicides can be selective and target broad-leaved species	
	5	Fungicide can be applied based on fungal disease forecasts **(Any 4)**	
	6	May be toxic to animals	
	7	May persist in the environment	
	8	May accumulate/be biomagnified in food chains	
	9	May lead to damage/imbalance in natural populations	
	10	May produce a selection pressure on a population	
	11	Could result in resistant populations **(Any 5)**	

29		Weeds:	8
	1	Weeds compete with/inhibit (crop) plants	
	2	Weeds reduce productivity/growth/yield	
	3	Annual weeds have rapid growth/short life-cycles/complete life-cycle within a year/produce many seeds/produce seeds with long-term viability	
	4	Perennial weeds have storage organs/vegetative reproduction	
	5	Pests: Pests eat/damage crops/plants/plant parts	
	6	Reduce productivity/growth/yield	
	7	Any from nematodes/insects/molluscs	
	8	Any other	
	9	Diseases: Diseases are caused by bacteria/fungi/viruses	
	10	Diseases are often spread/carried by invertebrates **(Any 8)**	
30	1	Weeds/pests/diseases can be controlled by cultural means	6
	2	Any linked example from ploughing/weeding/crop rotation/time of sowing	
	3	Selective weed killer/selective herbicide kills/affects certain plant species/broad-leaved weeds	
	4	Systemic weed killer spreads through the vascular system/phloem	
	5	Kills the whole plant/stops regrowth/regeneration/can kill pests which feed on plant	
	6	Applications of fungicide based on disease forecasts are more effective than treating diseased crop	
	7	Biological control is use of predator/parasite/disease of pest predator/parasite/control organism/disease	
	8	Integrated pest management (IPM) combines cultural, chemical and biological methods **(Any 6)**	

19   Animal welfare

What you need to know about behavioural indicators of poor welfare

There are costs, benefits and ethics of providing different levels of animal welfare in livestock production.

Intensive farming is less ethical than free-range farming due to poorer animal welfare.

Free range requires more land and is more labour intensive but can be sold at a higher price and animals have a better quality of life.

Intensive farming often creates conditions of poor animal welfare but is often more cost effective, generating higher profit as costs are low.

Behavioural indicators of poor animal welfare are **stereotypy**, **misdirected behaviour**, failure in sexual or parental behaviour and altered levels of activity (very low/apathy or very high/hysteria).

Key diagram

Behavioural indicators of poor welfare in livestock production: **(a)** stereotypy – repeated bar-chewing by pigs; **(b)** misdirected grooming by chickens leads to de-feathering; **(c)** altered activity levels shown by hysteric aggression in bulls; and **(d)** failure in parenting in sheep causing the abandoning of a lamb which has now been adopted by another ewe.

(a)

(b)

(c)

(d)

C-type questions

1. Give **one** example of a type of behaviour which could indicate poor animal welfare.   1

2. Name the type of behavioural indicator of poor animal welfare associated with repeated movements such as aimless pacing.   1

3. Name the type of behavioural indicator of poor animal welfare associated with inappropriate use of normal behaviour such as over-grooming of feathers by chickens.   1

A-type question

4. Describe issues that arise when deciding on the level of animal welfare to provide on a pig farm.   2

Extended response questions

5. Give an account of the costs, benefits and ethics of providing different levels of animal welfare in livestock production.   5

6. Give an account of the different behavioural indicators of poor welfare in livestock production.   5

Model answers and commentary

Question	Model answer	Marks	Commentary with hints and tips
1	Stereotypy; misdirected behaviour; failure in sexual or parental behaviour; altered levels of activity/very low (apathy)/very high (hysteria) levels of activity **(Any 1)**	1	These are the four examples given in the course specification.
2	Stereotypy	1	Repeated behaviour with no obvious goal or function.
3	Misdirected behaviour	1	Normal behaviour which is exaggerated or applied incorrectly.
4	Costs/expense; benefits/improved quality; ethical questions **(Any 2)**	2	Intensive farming is less ethical than free-range farming due to poorer animal welfare. Free range requires more land and is more labour intensive but can be sold at a higher price and animals have a better quality of life.

5	1	Free range requires more land	5
	2	Free range is more labour intensive	
	3	Free range produce can be sold at a higher price	
	4	The animals have a better quality of life	
	5	Intensive farming often creates conditions of poor animal welfare	
	6	Intensive farming is often more cost effective/generates higher profit/ has lower costs	
	7	Intensive farming is less ethical than free-range farming due to poorer animal welfare **(Any 5)**	
6	1	Stereotypy	5
	2	– Repeated aimless activity	
	3	Misdirected behaviour	
	4	– Normal behaviour but abnormally directed	
	5	Altered activity levels	
	6	– e.g. hysteria/apathy	
	7	Failure in sexual behaviour/parenting **(Any 5)**	

20 Symbiosis

What you need to know about symbiosis and parasitism

Symbiosis is a co-evolved intimate relationship between members of two different species.

Types of symbiotic relationship include parasitism and **mutualism**.

A **parasite** benefits in terms of energy or nutrients, whereas its **host** is harmed by the loss of these resources.

Parasites often have limited metabolism and cannot survive out of contact with a host.

Parasites are transmitted to new hosts using direct contact, **resistant stages** and **vectors**.

Some parasitic life-cycles involve **intermediate (secondary) hosts** to allow them to complete their life-cycle.

Key diagram

Examples of transmission of parasites to new hosts

Method of transmission	Meaning	Example
Direct contact	An infected organism passes parasites directly to another organism	A cat could pass fleas to another cat by brushing against its coat
Resistant stages	A parasite could form a resistant stage and be able to live without a host for a period of time	Some species of dog mite can exist away from their host in a protective covering which can be left in the environment then infect another dog
Vectors	Parasite can be actively carried from one host to another by a vector	A mosquito could bite a person infected with malaria parasites then pass the parasite on by biting a second person

C-type questions

1. Give the general name for an intimate co-evolved relationship between members of two different species.   1

2. Name the symbiotic relationship in which one species benefits and the other is harmed.   1

3. Give the meaning of the term 'parasite'.   1

4. State **one** way in which parasites are transmitted to new hosts.   1

5. Give the term which describes organisms which actively transfer
    parasites to hosts.   1

6. Give the term used for a third species in a parasite–host relationship in
    which the parasite can complete its life-cycle.   1

A-type questions

7. Describe the characteristics of a symbiotic relationship.   3

8. Explain the effect of a parasitic relationship on the host.   2

> **Many candidates have difficulty explaining the effect of a parasitic
> relationship on the host – remember to talk about the host losing
> energy or nutrients.**

9. Explain the benefit gained by some parasites in having an intermediate
    (secondary) host as part of their life-cycle.   2

> **Many candidates have difficulty describing the benefit of a secondary
> host to a parasite – remember to say that a secondary host allows the
> parasite to complete its life-cycle.**

➡ **Model answers and commentary can be found on page 153.**

What you need to know about mutualism

In mutualism, both mutualistic partner species benefit in an interdependent relationship.

Key diagram

An example of mutualism – the clownfish cleans the anemone, provides it with nutrients
from waste products and discourages predators of the anemone; the clownfish is protected
from its own predators by the stings of the anemone.

C-type question

10. Parasitism is one form of symbiosis.
    Name **one** other type of symbiosis.   1

A-type question

11. Explain why mutualism can be described as a type of symbiosis.   3

Extended response question

12. Give an account of symbiosis under the following headings:
    (a) parasitism   5
    (b) mutualism   2

Model answers and commentary

Question		Model answer	Marks	Commentary with hints and tips
1		Symbiosis	1	Many students fail to state that the relationship is between two **different species**.
2		Parasitism	1	The parasite gains and the host is harmed.
3		A parasite is an organism that benefits from its host in terms of gaining energy or nutrients	1	You must remember to include how the parasite benefits using the terms 'energy' and 'nutrients' and how the host is harmed by the loss of these resources.
4		Direct contact; resistant stages; vectors **(Any 1)**	1	A parasite vector is an organism that actively transfers the parasite from one host to another.
5		Vectors	1	Remember that secondary hosts do not actively transfer parasites but simply allow them to complete their life-cycle, whereas vectors are active agents of transmission.
6		Secondary/intermediate hosts	1	
7		Co-evolved; intimate relationship; between members of two different species	3	Quite a detailed and tricky definition to learn. The terms 'co-evolved' and 'intimate/close relationship' must be used. It must also be clear that this is between two different species.
8		Parasite benefits in terms of energy or nutrients; Host is harmed by the loss of these resources	2	Benefit, harm, energy and nutrients all need to be included to gain these marks.

9		Allows them to complete their life-cycle; Increases the probability/chance of infecting the primary host **OR** provides transport to the primary host **OR** supplies nutrients for the parasite's development until a suitable host is available	2	Remember that secondary hosts are not vectors.
10		Mutualism	1	Both mutualistic partner species benefit in an interdependent relationship.
11		It is an intimate; co-evolved relationship; between two different species in which both species in the interaction benefit	3	Need to give the definition of the symbiotic relationship and then give the meaning of mutualism.

12	(a)	1	Parasites gain energy/nutrients		5
		2	Host is harmed by loss of these resources		
		3	Parasites have more limited metabolism/cannot survive out of contact with host		
		4	Are transmitted by direct contact		
		5	Resistant stages		
		6	Vectors		
		7	Some have intermediate/secondary hosts **(Any 5)**		
	(b)	1	Mutualism benefits both partner species		2
		2	Partners are interdependent		
		3	Example – bacteria in herbivore gut/photosynthetic algae in coral polyps/clownfish and anemone (others possible) **(Any 2)**		

Social behaviour

What you need to know about social hierarchy, co-operative hunting and social defence

Many animals live in social groups and have behaviours that are adapted to group living such as **social hierarchy**, **co-operative hunting** and social defence.

Social hierarchy is a rank order within a group of animals consisting of **dominant** and **subordinate** members.

In a social hierarchy, dominant individuals carry out ritualistic (threat) displays while subordinate animals carry out **appeasement behaviour** to reduce conflict.

Social hierarchies increase the chance of the dominant animal's favourable genes being passed on to offspring. Animals often form **alliances** in social hierarchies to increase their social status within the group.

Co-operative hunting may benefit subordinate animals as well as dominant ones, as they may gain more food than by foraging alone.

In co-operative hunting, less energy is used per individual.

Co-operative hunting enables larger prey to be caught and increases the chance of success.

Social defence strategies increase the chances of survival as some individuals can watch for predators while others can forage for food.

In social defence, groups may adopt specialised formations when under attack, protecting their young.

Key diagram

Comparison of various social behaviours

Social behaviour	Species as an example	Survival value
Social hierarchy	Grey wolf	Lowers aggression and saves energy Experienced leadership guaranteed Most favourable genes likely to be passed on
Co-operative hunting	African wild dog	Increases chance of obtaining larger prey May gain more food than by foraging alone Subordinate individuals benefit Less energy is used per individual
Social defence	Musk ox	Early warning of predators can be given Younger individuals defended by group formation Predators intimidated or confused

C-type questions

1. Give a definition of the term 'co-operative hunting'.   1

2. Give the term used to describe a system of social organisation
    that results in a rank order of individuals.   1

3. Grey wolves live in packs in which a social hierarchy exists.
    The animals use co-operative hunting techniques.

    (a) Give a definition of the term 'dominance hierarchy'.   1

    (b) State **two** advantages to wolves of using co-operative hunting.   2

4. Give a survival benefit of social defence.   1

A-type questions

5. Explain the benefits of co-operative hunting strategies.   3

6. Explain the benefits of a social system involving a dominance hierarchy.   3

⟹ **Model answers and commentary can be found on page 159.**

What you need to know about altruism and kin selection

An **altruistic behaviour** harms the **donor** individual but benefits the **recipient**.

Behaviour that appears to be altruistic can be common between a donor and a recipient if they are related (kin).

The donor will benefit in **kin selection** in terms of the increased chances of survival of shared genes in the recipient's offspring or future offspring.

In reciprocal altruism, the roles of donor and recipient later reverse so that the original donor benefits.

Key diagram

(a) Worker bees caring for young as part of kin selection and **(b)** vampire bats at a roost where they may feed each other in reciprocal altruism

(a)   **(b)**

C-type questions

7. Give the meaning of the term 'altruism'.   1
8. Give the meaning of the term 'reciprocal altruism'.   1
9. Give the meaning of the term 'kin selection'.   1
10. Give the term that describes altruistic behaviour towards relatives.   1
11. On returning to their roost after feeding, vampire bats may regurgitate blood to feed an unrelated individual in the same social group.

    Give the term for this type of behaviour.   1

A-type questions

12. In honey bees, worker individuals are sterile but work to ensure that offspring of their relatives are fed.

    Explain the altruistic behaviour of the worker bees.   2
13. On returning to their roost after feeding, vampire bats may regurgitate blood to feed an unrelated individual in the same social group.

    Explain the survival value which could be gained by feeding an unrelated individual.   1

➡ **Model answers and commentary can be found on page 160.**

What you need to know about social insects

Social insects include bees, wasps, ants and termites.

Social insects have a society in which only some individuals (queens and drones) contribute reproductively.

Most members of a colony are sterile workers who co-operate with close relatives to raise relatives.

Other examples of workers' roles include defending the hive, collecting pollen and carrying out waggle dances to show the direction of food.

Sterile workers raise relatives to increase survival of shared genes.

Key diagram

The different types of individual in a honey bee colony – note that all of the individuals are closely related and so share many of their DNA sequences

Queen – reproductive female whose eggs produce all members of the colony

Drone – reproductive males who fertilise the queen's eggs. They are produced from unfertilised eggs of a queen

Worker – sterile females produced from fertilised eggs of a queen

C-type questions

14. Give **two** examples of social insects.   2
15. Give **two** examples of the roles of worker bees within their colony.   2

A-type question

16. Describe the reproductive contribution made by queens, drones and workers in a honey bee colony.   2

➡ **Model answers and commentary can be found on page 161.**

What you need to know about primate behaviour

Primates have a long period of parental care to allow learning of complex social behaviour.

Complex social behaviours support the social hierarchy.

Ritualistic display and appeasement behaviour reduce conflict.

Grooming, facial expression, body posture and sexual presentation are examples of appeasement behaviour.

Alliances form between individuals, which are often used to increase social status within the group.

➡ **There is more about social hierarchy in 'What you need to know about social hierarchy, co-operative hunting and social defence' earlier in this chapter.**

Key diagram

Social behaviour in primates – social grooming in a chimpanzee group can contribute to the formation of alliances within the group

C-type questions

17. Baboons are social primates which live in large groups. Within a group, individuals are ranked in a social hierarchy and have complex social behaviours. Some individual baboons form alliances with others in their group to increase their social status.

    Give **one** example of how increased social status can benefit an individual baboon.   1

18. Primates often use appeasement behaviour to reduce unnecessary conflict within the group.

    Give **one** example of this type of behaviour.   1

19. Give **one** form of behaviour shown by social primates to reduce unnecessary conflict.   1

20. Give **one** feature of parental care in primates which allows complex social behaviour to be learned.   1

A-type questions

21. Explain why long periods of parental care are needed for chimpanzee development.   1

22. Explain the advantage to chimpanzees of complex social behaviours such as the use of facial expressions.   1

 Many candidates have difficulty in describing a social hierarchy in primates and why these animals use ritualistic displays – remember that a social hierarchy is an adaptation for survival and ritualistic display helps maintain the hierarchy but avoids aggression and excessive use of energy.

Extended response questions

23. Write notes on social behaviour under the following headings:
    (a) altruism and kin selection   4
    (b) primate behaviour   5

24. Write notes on social behaviour in animals under the following headings:
    (a) social hierarchy and co-operative hunting   5
    (b) social insects   4

Models answers and commentary

Question		Model answer	Marks	Commentary with hints and tips
1		Hunting behaviour in which individuals work together to catch prey	1	Idea of a group 'working together' is essential.
2		Social/dominance hierarchy	1	A rank order of individuals within a group from dominant to subordinate members.

3 (a)		Dominance hierarchy is a rank order/pecking order of individuals in a social grouping of animals	1	Hierarchy = ranked according to relative status or authority.
(b)		Allows larger prey to be caught/kills to be made; increases the success rate of hunts; energy gained in food is greater than that lost in hunting **OR** Less energy used per individual **OR** They may gain more food than by foraging alone **(Any 2)**	2	**NOT** allows larger prey to be 'hunted' – must indicate prey being caught/killed. **NOT** 'saves energy' – must be less energy used **per individual**.
4		Early warning of predators can be given; younger individuals defended by group formation; predators intimidated or confused **(Any 1)**	1	Good examples of early warnings are in bird alarm calls and meerkat lookouts; stampeding zebra can confuse predators.
5		Co-operative hunting may benefit subordinate animals as well as dominant ones; as they may gain more food than by foraging alone; less energy is used per individual; co-operative hunting enables larger prey to be caught; and increases the chance of hunting success **(Any 3)**	3	**NOT** allows larger prey to be 'hunted' – must indicate prey being caught/killed. **NOT** 'saves energy' – must be less energy used **per individual**.
6		Increase the chances of the dominant animal's favourable genes being passed on to offspring; Animals often form alliances to increase their social status within the group; Lowers aggression/reduces conflict within the group which saves energy; Provides experienced leadership **(Any 3)**	3	Good examples of a social hierarchy are found in lions, gorillas and hunting dogs.
7		Behaviour that harms the donor individual but benefits the recipient	1	The terms 'donor' and 'recipient' must be included in your answer.

8		Where the roles of the donor and recipient later reverse	1	Reciprocal = done in return.
9		Altruistic behaviour between related individuals **OR** Organisms donating resources to those with whom they share genetic material	1	Kin = family and relatives.
10		Kin selection	1	The donor benefits in terms of the increased chances of survival of shared genes in the recipient's offspring or future offspring.
11		Reciprocal altruism	1	The reference to 'unrelated individuals' means that this form of altruism is **not** kin selection but reciprocal.
12		Worker bees feed the offspring of relatives because they have shared genes; The feeding helps ensure that the offspring survive	2	Behaviour harms the donor individual but benefits the recipient.
13		Explanation to cover reciprocal altruism – There is an expectation that the other organism will act in a similar manner at a later time **OR** By feeding an unrelated individual, this gesture may be reciprocated/paid back on another occasion giving the individual a food source and increasing survival	1	This question referred to survival value and so it would not be sufficient to simply state that it is a gesture that is given back.
14		Bees; wasps; ants; termites **(Any 2)**	2	These are the social insects mentioned in the course specification.
15		Defend the hive; collect pollen; perform waggle dances to show direction of food **(Any 2)**	2	Remember that these behaviours refer to honey bees.
16		Only queens and drones are fertile/reproduce; Workers co-operate to raise the offspring of relatives/ queens/drones	2	The caste names in social insects must be known.

17			Higher-ranking females appear to have access to better food; and have a higher infant survival rate than their low-ranking counterparts **(Any 1)**	1	Remember that alliances are formed between individuals to increase their social status.	
18			Facial expressions; body postures; sexual presentation **OR** Example – eye closing/aversion/not baring teeth; lowering of body position/bowing actions; the presentation of genitalia by females to males **(Any 1)**	1	Conflict and aggression uses energy and could result in injury, which would reduce the effectiveness of the group.	
19			Ritualistic display **OR** specific example; appeasement behaviour **OR** specific example; general examples – grooming/facial expression/body posture/sexual presentation **(Any 1)**	1	Reduction of conflict also saves energy.	
20			Primate offspring have a long period of parental care	1	Good examples of learned behaviours include ritualistic display and appeasement.	
21			It allows the opportunity/time to learn complex social behaviours	1		
22			It helps support the social structure of the chimps/species/group	1	The social structure increases survival.	
23	(a)	1	Altruism:			4
			Donor harmed			
		2	Recipient benefits			
		3	Example of altruism			
		4	Roles can be reversed during reciprocal altruism **(Any 3)**			
		5	Kin selection:			
			Kin are close relatives who share genes with recipient			
		6	Kin selection is donating resources to kin **(Any 1)**			

	(b)	1	Primates are monkeys, apes and humans	5
		2	They have complex social behaviour	
		3	Born helpless and have long dependency period	
		4	Ritualistic **OR** appeasement behaviour designed to reduce conflict	
		5	Examples of behaviour with description from: grooming/mutual preening; facial expression/body posture/sexual presentation	
		6	Another example from the list **(Any 5)**	
24	(a)	1	Social hierarchy is a rank order/pecking order in a group of animals **OR** dominant/alpha **AND** subordinates/lower rank	5
		2	Aggression/fighting/conflict/violence reduced	
		3	Ritualistic display/appeasement/threat/submissive behaviour **OR** alliances formed to increase social status **OR** descriptive examples	
		4	Ensures best/successful genes/characteristics are passed on **OR** guarantees experienced leadership **(Any 3)**	
		5	Co-operative hunting is where animals hunt together to obtain food	
		6	Increases hunting success **OR** allows larger prey to be brought down/ hunted and killed **OR** more successful than hunting individually	
		7	Subordinate animals all get more food/energy than hunting alone **OR** less energy used/lost per individual **(Any 2)**	
	(b)	1	Any two examples – bees, wasps, ants, termites	4
		2	Only some members of colony/hive reproduce/are fertile **OR** queen **AND** males/drones mate/reproduce **OR** only queen lays eggs **OR** some/most members of colony are sterile/infertile/do not reproduce **OR** some/most of colony are workers who are sterile	
		3	Examples of worker roles, **ANY** one from: raise relatives, defend the hive, collect pollen/nectar/food, waggle dance to show direction of food, etc.	
		4	Social insects show kin selection/altruism between related individuals	
		5	Increases/helps survival of shared genes **OR** so shared genes are passed on to next generation **(Any 4)**	

CHAPTER 22

Components of biodiversity

What you need to know about the components of biodiversity

Components of biodiversity are **genetic diversity**, **species diversity** and **ecosystem diversity**.

Genetic diversity is the number and frequency of all the alleles within a population.

If one population of a species dies out then the species may have lost some of its genetic diversity, and this may limit its ability to adapt to changing conditions.

Species diversity comprises the number of different species in an ecosystem (the **species richness**) and the proportion of each species in the ecosystem (the **relative abundance**).

A community with a **dominant species** has a lower species diversity than one with the same species richness but no particularly dominant species.

Ecosystem diversity refers to the number of distinct ecosystems within a defined area.

➡ **Threats to biodiversity are covered in Chapter 23.**

Key diagram

Measurable components of biodiversity

Ecosystem diversity refers to the number of distinct ecosystems within a defined area

The number of different species in an ecosystem (the species richness) and the proportion of each species in the ecosystem (the relative abundance)

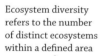

Ecosystem diversity   Species diversity

Genetic diversity

Genetic diversity comprises the genetic variation represented by the number and frequency of all the alleles in a population

Many candidates have difficulty in describing genetic diversity – remember to talk about the number and frequency of alleles in a population.

C-type questions

1. Give the meaning of the term 'genetic diversity'.   1
2. Give the term used to describe the number and frequency of alleles in a population.   1
3. Give the meaning of the term 'species diversity'.   2
4. Give the term used to describe the measure of species richness and relative abundance.   1
5. Give the meaning of the term 'ecosystem diversity'.   1
6. Give the meaning of the term 'species richness'.   1
7. Give the term used to describe the number of different species in a community.   1
8. Give **one** component of genetic diversity.   1

A-type questions

9. Describe the effect of the presence of a dominant species on the species diversity of an ecosystem.   2
10. Explain the negative impact on a species resulting from the loss of many individuals of a population.   2

Extended response question

11. Give an account of biodiversity and its measurement with reference to genetic diversity, species diversity and ecosystem diversity.   5

Model answers and commentary

Question		Model answer	Marks	Commentary with hints and tips
1		The number **OR** frequency of all the alleles within a population	1	Low genetic diversity limits the ability of a species to adapt to changing conditions.
2		Genetic diversity	1	Think genes – think alleles (number and frequency).
3		The number of different species in an ecosystem/the species richness; and the proportion of each species in the ecosystem/the relative abundance	2	Dominant species reduces species diversity but not species richness.
4		Species diversity	1	Think species – think number and abundance.
5		The number of distinct ecosystems within a defined area	1	Ecosystems range from tropical rainforests down to pond ecosystems.

6		The number of different species in an ecosystem	1	Remember that the other component of species diversity is the relative abundance of the different species present.
7		Species richness	1	
8		The number/frequency of all the alleles within a population	1	Populations with very low genetic diversity show the bottleneck effect.
9		It has a lower species diversity; because although the species richness may be high the relative abundance of some species is low	2	Dominant species are so abundant that the populations of other species are low making them vulnerable to local extinction.
10		The species may lose some of its genetic diversity; and this may limit its ability to adapt to changing conditions	2	Loss of alleles reduces adaptive potential.

11	1	Genetic diversity is the number of alleles in a population	5
	2	And frequencies of alleles in a population	
	3	Loss of a population can result in loss of genetic diversity of a species **(Any 2)**	
	4	Number of different species is species richness	
	5	Relative abundance is the proportion of each species present	
	6	A dominant species reduces species diversity	
	7	Ecosystem diversity is the number of distinct types of ecosystem in a defined area **(Any 3)**	

23 Threats to biodiversity

What you need to know about exploitation and recovery

With over-exploitation, populations can be reduced to a low level but may still recover.

Some species have a naturally low genetic diversity in their population and yet remain viable.

The **bottleneck effect** relates to small populations that may lose the genetic variation necessary to enable evolutionary responses to environmental change.

In small populations, this loss of genetic diversity can be critical for many species, as inbreeding can result in poor reproductive rates.

Key diagram

Model to represent the changes in genetic diversity which cause the bottleneck effect

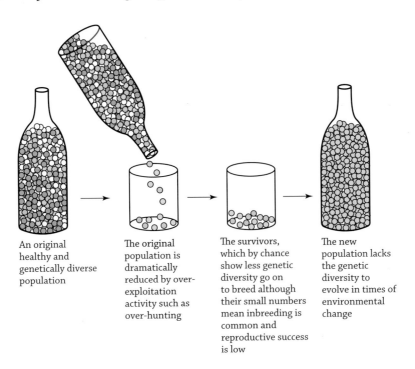

An original healthy and genetically diverse population

The original population is dramatically reduced by over-exploitation activity such as over-hunting

The survivors, which by chance show less genetic diversity go on to breed although their small numbers mean inbreeding is common and reproductive success is low

The new population lacks the genetic diversity to evolve in times of environmental change

C-type question

1. Give **one** effect of inbreeding in small populations.   1

➡ **There is more about inbreeding in Chapter 17.**

A-type questions

2. Describe what is meant by the bottleneck effect.   1
3. Describe the effect of over-exploitation and later recovery on
    the genetic diversity of a species.   2
4. Explain the possible impact of low genetic diversity in a small population.   2

➡ **Model answers and commentary can be found on page 171.**

What you need to know about habitat loss

The clearing of habitats has led to habitat fragmentation.

Degradation of the edges of **habitat fragments** results in increased competition between species as the fragment becomes smaller, often resulting in a decrease in biodiversity.

More isolated fragments and smaller fragments exhibit lower species diversity.

To remedy widespread habitat fragmentation, isolated fragments can be linked with **habitat corridors**.

Habitat corridors allow movement of animals between fragments, increasing access to food and choice of mate.

Habitat corridors may lead to recolonisation of small fragments after local extinctions.

Key diagram

Representation of habitat fragmentation and conservation by the creation of habitat corridors

Original habitat

Fragmentation

Conservation

more degradation occurs at habitat edges

smaller, more isolated fragment

habitat corridor

Large block of habitat with intact edges and high biodiversity

Clearing of habitat leads to habitat fragmentation. Degradation of the edges of habitat fragments results in increased competition between species as the fragment becomes smaller. More isolated and smaller fragments have a lower species diversity.

Fragments connected by habitat corridors by which isolated fragments can be linked to allow movement of animals between fragments increasing access to food and choice of mate. This may lead to recolonisation of small fragments after local extinctions.

C-type questions

5. State how habitats become fragmented.   1
6. Describe the effect of degradation of habitat fragment edges on species.   2
7. State what is meant by the term 'habitat corridor'.   2

A-type questions

8. Compare the species diversity in small isolated fragments compared with
    larger ones.   2
9. Habitat corridors can be created to remedy habitat fragmentation.

    Explain how a habitat corridor can increase biodiversity after local extinction.   2

➡ **Model answers and commentary can be found on page 171.**

What you need to know about introduced, naturalised and invasive species

Introduced (non-native) species are those that humans have moved either intentionally or accidentally to new geographic locations.

Those that become established within wild communities are termed **naturalised species**.

Invasive species are naturalised species that spread rapidly and eliminate native species, therefore reducing species diversity.

Invasive species may well be free of the predators, parasites, pathogens and competitors that limit their population in their native habitat.

Invasive species may prey on native species, outcompete them for resources or hybridise with them.

Key diagram

Example of how an introduced species may become invasive

Species	Definition	Example
Introduced (non-native)	Species moved intentionally or accidentally by humans to new geographic locations	In the 1870s grey squirrels were intentionally **introduced** to the UK from north America as ornamental species on large estates
Naturalised	Introduced species which have become established in wild communities	The grey squirrel became established in wild woodland communities in parts of the UK
Invasive	Naturalised species which spread rapidly and eliminate native species from communities by predation, outcompeting them or hybridising with them	Grey squirrels started to outcompete the native red squirrel in many parts of the UK and have now been classed as **invasive**, posing a threat to biodiversity

C-type questions

10. State what is meant by an introduced species.   1

11. State what is meant by a naturalised species.   1

12. State what is meant by an invasive species.   1

13. Give **one** reason why the population of an invasive species may increase at the expense of the native species.   1

14. Give the term used for a naturalised species which eliminates native species.   1

15. Give the name applied to introduced species which have become established in wild communities.   1

16. Give **two** types of damage which large populations of invasive species can inflict on the biodiversity of an ecosystem in which they have become established.   2

A-type question

17. Explain why the population of invasive species can become very large in their new habitats.   2

Extended response questions

18. Give an account of habitat fragmentation and measures that can be taken to minimise its effects.   5

19. Write notes on invasive species.   4

20. Discuss introduced species and their impact on native populations.   5

Many candidates have difficulty in describing naturalised species – remember to use the phrase 'become established in wild communities'.

Many candidates have difficulty in defining an invasive species and explaining how an invasive species can be identified – remember that they may not have natural predators, parasites or pathogens.

Model answers and commentary

Question		Model answer	Marks	Commentary with hints and tips
1		Small populations may lose the genetic variation necessary to enable evolutionary responses to environmental change (the bottleneck effect) **OR** Inbreeding in small populations can result in poor reproductive success/rates	1	Small populations are forced to inbreed and this can lead to inbreeding depression.
2		When small populations lose the genetic variation necessary to enable evolutionary responses to environmental change	1	Good examples include the Northern elephant seal, cheetahs and European bison.
3		Small populations lose the genetic variation necessary to enable evolutionary responses to environmental change; following recovery, the population has less genetic diversity than the original population due to inbreeding	2	Good examples of this effect include Northern elephant seal, European bison, Hawaiian goose and Atlantic cod.
4		Loss of the genetic variation necessary to enable evolutionary responses to environmental change; poor reproductive success/rates	2	
5		Habitats are cleared	1	Clearing might mean deforestation, the draining of wetlands or urbanisation.
6		Increases competition between species; Results in decrease in biodiversity	2	Degradation of edges forces species away from edges and into competition with interior species.
7		Habitat created to connect habitat fragments; Separated by habitat clearance	2	Remember a corridor allows individuals to meet, mate and repopulate and increase/restore biodiversity following local extinction.
8		Less biodiversity in small fragments; Less biodiversity in more isolated fragments	2	Smaller fragments have lower species richness and the relative abundance could be changed as well.

9		Corridors allow movement of animals between fragments; increasing access to food/choice of mate; lead to recolonisation of small fragments **(Any 2)**	2	A good example of a habitat corridor is the tiger corridor through Bhutan.
10		A species moved intentionally or accidentally to a new geographical location	1	Many examples in the UK, such as grey squirrels and rhododendron.
11		Species that have become established within wild communities	1	It is probably best to use this exact phrase.
12		Invasive species are naturalised species that spread rapidly and eliminate native species	1	Their spread may be due to the absence of predators, parasites or pathogens.
13		Invasive species may be free of predators/parasites/pathogens/ competitors	1	Remember that these factors limit their population in their native habitat.
14		Invasive species	1	A good example of an invasive species which preys on native species is the cane toad introduced into Australia.
15		Naturalised species	1	Some introduced species become naturalised but do not become invasive – a good example is the Mandarin duck in the UK.
16		Invasive species may prey on native species; outcompete them for resources; hybridise with them **(Any 2)**	2	When invasive species hybridise with native species, the hybrids are infertile and populations can be reduced.
17		Invasive species may well be free of the predators; parasites; pathogens; competitors that limit their population in their native habitat **(Any 2)**	2	Very similar to Question 13 but more examples are needed for both marks.

18	1	Habitat fragmentation occurs when habitat is split into small parts	5
	2	Fragments support lower species richness than a large area of the same habitat	
	3	Fragments may be degraded at their edges	
	4	Edge species may invade interiors and displace other species **(Any 3)**	
	5	Isolated fragments can be connected by habitat corridors	
	6	Species can feed/find mates within corridors	
	7	Recolonisation of deserted fragments can occur **(Any 2)**	
19	1	Introduced species **OR** species moved intentionally/accidentally by humans	4
	2	These species become naturalised species when they are established (in the new area)	
	3	Spread rapidly and may eliminate/kill off/destroy native/indigenous/original species	
	4	Prey on/outcompete/hybridise with native/indigenous/original species	
	5	Natural/original predators/parasites/pathogens/competitors are not present in new area **(Any 4)**	
20	1	Introduced species are those that humans have moved intentionally/accidentally to new geographic locations	5
	2	Those species that become established within wild communities are termed naturalised species	
	3	Invasive species are naturalised species that spread rapidly and eliminate native species	
	4	Invasive species may be free of the predators/parasites/pathogens/competitors that limit their population in their native habitat	
	5	Introduced species may prey on native species	
	6	Introduced species may outcompete native species for resources **(Any 5)**	

Assignment

Introduction

The assignment has a total mark allocation of 20 marks. This is scaled to 30 marks to represent 20% of the overall marks for the course assessment. The remaining 80% is allocated to the question paper (examination). No more than 8 hours should be spent on the whole assignment.

Candidates research and report on a topic that allows them to apply skills and knowledge in biology at a level appropriate to Higher. The topic must be chosen with guidance from teachers and/or lecturers and must involve experimental work. The assignment is an individually produced piece of work started at an appropriate point in the course and conducted under controlled conditions.

Structure of the assignment

The assignment has two stages.

Research stage

The research stage must involve **experimental work** which allows measurements to be made. Candidates must also gather **data/information from the internet, books or journals**. The research stage should take around 6 hours.

In the research stage, teachers and/or lecturers must agree the choice topic with the candidate, provide advice on the suitability of the candidate's aim and can supply instructions for the experimental procedure. Candidates must undertake research using websites, journals and/or books and a wide list of URLs and/or a wide range of books and journals may be provided. Teachers and/or lecturers must **not** provide an aim, experimental data, a blank or pre-populated table for experimental results or feedback on their research.

Report stage

Candidates must produce a report on their research. The report stage should take no more than 2 hours. The report is submitted to SQA for external marking. The only materials which can be used in the report stage are SQA instructions for candidates, the candidate's raw experimental data, data/information taken from the internet or literature with a record of these source(s) and extract(s) from the internet/literature sources to support the underlying biology and the experimental method, if appropriate. There is no word count.

Mark allocations

The table below gives details of the mark allocation for each section of the report.

Section	Expected response	Marks
Aim	An aim that describes clearly the purpose of the investigation.	1
Underlying biology	An account of biology relevant to the aim of the investigation.	4
Data collection and handling	A brief summary of the approach used to collect experimental data.	1
	Sufficient raw data from the candidate's experiment.	1
	Data, including mean values, presented in a correctly produced table.	1
	Data/information relevant to the experiment obtained from an internet/literature source.	1
	A citation and reference for a source of internet/literature data or information.	1
Graphical presentation	An appropriate format from the options of line graph or bar graph.	1
	The axes of the graph have suitable scales.	1
	The axes of the graph have suitable labels and units.	1
	Data points are plotted accurately with a line or clear bar tops (as appropriate).	1
Analysis	A correct comparison of the experimental data with data/information from the internet/literature source or a correctly completed calculation(s) based on the experimental data, linked to the aim.	1
Conclusion	A valid conclusion that relates to the aim and is supported by all the data in the report.	1
Evaluation	Evaluation of the investigation.	3
Structure	A clear and concise report with an informative title.	1
TOTAL		**20**

Tips on challenging aspects of the assignment

Aim
You must give an aim which clearly states an independent and a dependent variable which then allows you to draw a valid conclusion.

Underlying biology
You must choose a topic directly related to the Higher Biology course and be sure to provide knowledge at Higher level – not N5.

Select information
You must ensure that both of your sources include data and are not simply underlying knowledge.

Data collection and handling
You must only provide a brief summary of the approach you used to collect experimental data – not a detailed account.

Graphical presentation
You must transfer complete table headings, including units, to graph labels.

You must select an appropriate scale so that your graph occupies at least half of the graph paper you are using.

Analysis
You must describe the relationship shown by the selected or processed data clearly.

Conclusion(s)
You must give a concise conclusion related to the aim and not an account of the entire investigation. You must be careful to take account of both your experimental data and your second source of data when drawing a conclusion, even if the data is conflicting.

Evaluation
Ensure that you use the terms 'robust', 'reliable' and 'valid' correctly.

Structure
You must give full references for all sources you used.

You must also give the title and aim for your experiment.